The Broken Image

The Broken Image

Man, Science and Society

by Floyd W. Matson

GEORGE BRAZILLER • NEW YORK

For Carla

The present book might be described as a report on the encounter between the "two cultures" (the scientific and the humane) at the point of its greatest impact: that is, in the field of the social sciences. But it makes no claim to a detached impartiality; it is animated by an active concern and a definite commitment. The nature of that commitment may be expressed in the hope that, when a *rapprochement* is ultimately achieved between the opposing cultures, it will take the form not of a "scientific humanism" but of a humane science. There is a difference. Despite the contrary inclinations of C. P. Snow and others, it is today less urgent that the humanities should become imbued with the values of science than that science should become alert to the values of humanity. To put it only a little differently, our most pressing educational and cultural need is not for the indoctrination of men in the directives of science, but (to echo the plea of Gerard Piel) for the enlistment of science in the cause of man.

And, more pointedly, for the enlistment of *social* science. For the human self-image which finds reflection in those capacious and definitive "mirrors for man," the social studies, has traditionally been filtered through a screen of metaphors borrowed from the neighboring laboratories of the physical sciences. The assumption has been, of course, that the resultant portrait is thereby purified and clarified—purged alike of the burdensome excrescences of metaphysics and the artificial colorations of romantic vanity. In a word, the process is objective. But there, precisely, is the rub. "Man," as Paul Tillich has warned us, "resists objectification; and if his resistance to it is broken, man himself is broken."

It is the primary thesis of this book, then, that the historic reliance of the social sciences upon root metaphors and routine methods appropriated from classical mechanics has eclipsed the ancestral liberal vision of "the whole man, man in person" (to use Lewis Mumford's phrase)—and has given us instead a radically broken self-image. The tragic history of the breaking of the human image parallels the disintegration of the inner sense of identity, the flight from autonomous conduct to automaton behavior, in the modern world.

But the story need not, and in fact does not, end there. Our own century has witnessed the rise of an opposite and potentially equal force, a positive countermovement whose wellspring is in natural science itself and whose fundamental motive is a passionate concern with the mending of the broken image and the recovery of an integral vision. Out of the wreckage of displaced allegiances and abandoned human values which the mechanization of social science has left in its wake, this insurgent movement of reconstruction has steadily been taking form and gaining voice. Its advent has not, to be sure, passed unnoticed; in each of the special fields in which it has found expression it is still the subject of active controversy. And yet, as a general movement of thought, it has remained curiously unacknowledged and widely unappreciated with respect to the revolutionary changes it portends for the proper study of mankind—for, that is to say, what man may make of man.

The present volume begins with a telescopic overview of the penetration and transformation of social thought, over nearly three centuries, by the world view of modern science: the legacy of Galileo and Newton, reinforced in the last century by the impact of Darwin. The extent and significance of this social scientism in our own age is then explored through a critique of that contemporary paradigm of a "natural science" of man and society—behaviorism—in both its psychological and political expressions. The chapters of Part II, in turn, undertake to trace the origins and career of the affirmative countermovement in postmodern science—from its sources in the new physics of uncertainty and complementarity, through its consensual endorsement in the biology of freedom, to its several distinctive formulations in those adventurous outposts of psychology which seek to dissolve the barriers of detachment and disinterest between observing man and man observed, and so to bring about a genuine meeting of minds. The final chapter, a concluding scientific postscript, attempts to place this complementary perspective—with its new image of indivisible man—within the proper frame of reference, at once classic and contemporary, in the science of political understanding.

In the writing of this book I have happily accumulated an immense burden of indebtedness, both tangible and intangible. My indirect obligations, too numerous to specify, are to all those great advocates of the human dialogue, from Socrates to Martin Buber, upon whose wisdom I have traded and whose insights have shaped my own perspective. Their identities, however, will not remain a mystery to the reader of these pages.

Among my specifiable debts, the greatest is to four Berkeley colleagues of the University of California whose wise and tolerant guidance was indispensable to the launching of the project which became this book, and to its successful navigation through waters not always placid: Professors Eugene Burdick, Peter H. Odegard and Jacobus tenBroek, all of the Department of Political Science; and Professor Victor F. Lenzen, of the Department of Physics.

For sustained support and counsel—and simple inspiration—I am deeply obligated to four other distinguished scholars: Professors Gordon W. Allport, William Ernest Hocking, Ashley Montagu and Lewis Mumford.

To Dr. Robert M. Hutchins, Dr. Scott Buchanan, Hallock Hoffman and the other resident members of the Center for the Study of Democratic Institutions, I am grateful for two productive opportunities during the summer of 1961 to discuss the themes of the book in the exemplary intellectual atmosphere of the Santa Barbara Center.

To all of the following scholars, who have read portions of the work in manuscript, I am thankful for constructive comments and suggestions: Hannah Arendt, Jacques Barzun, Herbert Blumer, the late P. W. Bridgman, Giorgio de Santillana, William Y. Elliott, Jacob Freid, Don Geiger, Hans H. Gerth, Kurt Goldstein, Howard Gossage, F. A. Hayek, Albert H. Hobbs, Terry Hoy, Alexandre

Koyré, Joseph Wood Krutch, Leo Lowenthal, Wallace A. Mac-Donald, Edward F. Meylan, Barrington Moore, Jr., Hans J. Morgenthau, Albert D. Mott, Herbert J. Muller, Gardner Murphy, Robert A. Nisbet, J. Robert Oppenheimer, Wolfgang K. H. Panofsky, William G. Pollard, Philip Rieff, John P. Roche, Paul A. Schilpp, Edward Teller, Paul Tillich, Nicholas S. Timasheff, and Robert T. Whalen.

I owe a particular debt of appreciation to Mr. Edwin Seaver, Editor-in-Chief of George Braziller, Inc., for a truly creative job of editing as well as for the stimulus of his unflagging interest and encouragement. And for heroic assistance far beyond the call of kinship, I wish to record my gratitude to my mother, Mrs. Gwendolyn Matson.

FLOYD W. MATSON

Berkeley, October, 1963.

Acknowledgments

For permission to quote from copyright material, thanks are due to the following:

Appleton-Century-Crofts—for passages from Clark L. Hull, *Principles of Behavior*. Copyright, 1943, D. Appleton-Century Company, Inc.

Basic Books, Inc.—for passages from Ernest G. Schachtel, *Metamorphosis* (New York: Basic Books, 1959).

Harcourt, Brace and World, Inc.—for passages from Werner Heisenberg, *The Physicist's Conception of Nature* (New York: Harcourt, Brace, 1958); and Karl Mannheim, *Ideology and Utopia* (New York: Harcourt, Brace, 1949).

Harper and Row, Publishers, Inc.—for passages from *Job Horizons: A Study of Job Satisfaction and Labor Mobility* by Lloyd G. Reynolds and Joseph Schister, Yale Labor and Management Series (New York: Harper, 1949); and Carl J. Friedrich, *Constitutional Government and Politics* (New York: Harper, 1937).

Harvard University Press—for passages from J. Bronowski, *The Common Sense of Science* (Cambridge: Harvard University Press, 1955), and A. C. Crombie, *Medieval and Early Modern Science*, Vol. II (Cambridge: Harvard University Press).

The Humanities Press, Inc.—for passages from E. A. Burtt, *The Metaphysical Foundations of Modern Physical Science* (New York: Humanities Press, 1950).

Alfred A. Knopf, Inc.—for passages from *The Thomas Mann Reader*, ed. by Joseph Warner Angell (New York: Knopf, 1950).

McGraw-Hill Book Company, Inc.—for passages from Hans von Hentig, *Crime: Causes and Conditions*. Copyright, 1947. McGraw-Hill Book Company. Used by permission.

David McKay Company—for passages from George A. Lundberg, *Can Science Save Us?* (New York: McKay, 1947).

The Macmillan Co.—for passages from B. F. Skinner, *Science and Human Behavior* (New York: Macmillan, 1953); A. N. Whitehead, *Modes of Thought* (New York: Macmillan, 1938); and Sir James Frazer, *The Golden Bough* (New York: Macmillan, one-vol. ed., 1951).

The New American Library—for passages from *Great Dialogues of Plato*, tr. by W. H. D. Rouse, ed. by Eric H. Warmington and Philip G. Rouse. (C) 1956 by John Clive Graves Rouse (New York: Mentor Books, 1956).

W. W. Norton & Co., Inc.—for passages from Karen Horney, *Neurosis and Human Growth* (New York: Norton, 1950).

Prentice-Hall, Inc.—for passages from Pauline V. Young, *Scientific Social Surveys and Research*. (C) 1956. By permission of Prentice-Hall, Inc., Englewood Cliffs, N.J.

Simon and Schuster, Inc.—for passages from J. E. Oppenheimer, *The Open Mind* (New York: Simon and Schuster, 1955).

The University of Chicago Press—for passages from Michael Polanyi, *The Logic of Liberty* (Chicago, 1951); and *The State of the Social Sciences*, ed. by Leonard D. White (Chicago, 1956).

C. A. Watts & Co., Ltd.—for passages from Ludwig von Bertalanffy, *Problems of Life* (London, 1952).

Wesleyan University Press—for passages from Norman O. Brown, *Life Against Death*. Copyright (C) 1959 by Wesleyan University. Reprinted by permission of Wesleyan University Press.

The Yale University Press—for passages from Hans Spemann, *Embryonic Development and Induction* (New Haven, 1938); Ernst Cassirer, *Determinism and Indeterminism in Modern Physics* (New Haven, 1956); and George Gaylord Simpson, *The Meaning of Evolution* (New Haven, 1949).

Contents

xiii

PART TWO / HUMANIZATION—FROM PHYSICS
TO POLITICS

The Broken Image

"Not the Power Man, not the Profit Man, not the Mechanical Man, but the Whole Man, Man in Person, so to say, must be the central actor in the new drama of civilization. . . . If technics is not to play a wholly destructive part in the future of Western Civilization we must now ask ourselves, for the first time, what sort of society and what kind of man are we seeking to produce?"

—Lewis Mumford, *In the Name of Sanity* (1954)

Part One

The Great Machine

CHAPTER I

The Mechanization of Man: Man, Science and Society—an Overview

Homo ex Machina: The Legacy of Newton

THE SCIENTIFIC WORLD VIEW which has been identified for three centuries with the names of Galileo and Newton—the cosmology of classical mechanics—looked upon an infinite universe of perfect symmetry and absolute precision. It was, in fact, nothing less than the image of the Great Machine. All that happened on earth and in the heavens, as J. Robert Oppenheimer has written of this cosmic vision, had its natural and knowable efficient cause. "The great machine had a determinate course. A knowledge of its present and therefore its future for all time was, in principle, man's to obtain, and perhaps in practice as well." And something more: "The giant machine was not only causal and determinate; it was *objective* in the sense that no human act or intervention qualified its behavior."[1]

This was the essential characteristic of the vast perpetual-motion apparatus conceived by Descartes and perfected by Newton: it was not only untouched and untouchable by human hands, but devoid of all purpose. It may be that the most revolutionary achievement of the Newtonian world view, as Niels Bohr among

19

others has suggested, was this expulsion of purpose from the universe.[2] To be sure, the task had been begun a century earlier by Copernicus, whose own subversive astronomy led such devout souls as John Donne to tremble for the newly displaced persons of earth:

> ... [the] new philosophy calls all in doubt,
> The Element of fire is quite put out;
> The Sun is lost, and th'earth, and no mans wit
> Can well direct him where to looke for it.
> 'Tis all in peeces, all cohaerence gone;
> All just supply, and all Relation.[3]

The Copernican revolution had dislodged man from the center of the universe; it remained for the Galilean-Newtonian revolution to remove him from the universe altogether.[4] Through the inexorable reduction of all knowable reality to the dimensions of objective mechanism, the gap between the knower and the known, between the subjective self and the world, came to be the measure of the distance between appearance and reality. Only the *primary* qualities (number, figure, magnitude, position and motion), inhering in the object "out there," were henceforth to be regarded as substantially real; the *secondary* qualities (all else which the senses perceive or the mind assembles), inhering in the human subject, were in effect unreal. "I think," said Galileo, "that these tastes, odours, colours, etc., . . . are nothing else than mere names, but hold their residence solely in the sensitive body; so that if the animal were removed, every such quality would be abolished and annihilated."[5]

And so, for purposes of science, the animal *was* removed —except as insensitive body, or more accurately as mechanism.[6] The consequences of this displacement have not yet, after three centuries, fully run their course. The "infinitely closed" universe of Newtonian cosmology seemed to seal man's fate by abolishing man's hope; and not only his hope but (in swift succession) his spiritual sovereignty over the natural world, his autonomy apart from it, and his distinctive reality within it. The authority of Newton, as E. A. Burtt has rather mercilessly expressed it,

> was squarely behind that view of the cosmos which saw in man a puny irrelevant spectator (so far as a being wholly imprisoned in a dark room can be called such) of the vast mathematical system whose regular motions according to mechanical principles constituted the world of nature. . . . The world that people had thought themselves living in—a world rich with colour and sound, redolent with fragrance, filled with gladness, love and beauty, speaking everywhere of purposive harmony and creative ideals— was crowded now into minute corners in the brains of scattered organic beings. The really important world outside was a world hard, cold, colourless, silent and dead. . . .[7]

These were not, of course, the inferences immediately and everywhere drawn from the "corpuscular philosophy" of universal mechanism. The traumatic effect of what Alexander Koyré has called "the scientific and philosophical destruction of the cosmos"[8] came upon most men not as a shock of recognition, counseling humility and curbing pride, but rather as a flash of inspiration and the vision of a new and greater mastery. The seventeenth century was already an age of industrious ideology and enterprise; to the troublesome question, "What profit a man . . . ?", there were more than a few prepared to answer (if at first only in a whisper) that the profits might be great indeed, were men to possess the universal knowledge which could unlock the great machine and expose its secret mainsprings. Even while Nature and Nature's laws still lay hid in night, Francis Bacon had looked forward to a time when "Natural Magick" would achieve the transmutation of the elements.[9] In the generation of Newton, when all was light, the belief in science swiftly became the faith of Scientism—the magical conception of natural science as omniscient and omnipotent.

It was a scientist of our own century who observed that modern man may have succeeded in emancipating himself from his belief in the magical powers of supernatural agencies only to plunge into an equally naive commitment to the magical powers of science—"a belief that precise measurement and prodigious calculation will lead not only to widespread human happiness, . . . but to a knowledge of ultimate reality, which the philosophers have vainly sought through the ages."[10] If the crime of Galileo was merely to intimate such a possibility, the dream of Descartes was to bring it to full and vivid reality. On a November night in the year 1619, Descartes experienced a fateful dream in which (as it seemed to him on later reflection) the Spirit of Truth opened to his gaze the treasure of all the sciences, wherein "the human mind played no part," and revealed to the philosopher "the foundations of the *admirable Science*" (*mirabilis scientia fundamenta*).[11] The fruit of this nocturnal vision was the famous *Discourse on Method*, whose title was originally conceived by Descartes as the "Project of a Universal Science Destined to Raise Our Nature to Its Highest Degree of Perfection." It was no mere ordinary science which had been revealed to him by the Spirit, but nothing less than the Science of Sciences—the universal mechanics underlying and illuminating all reality. To Descartes from that moment on, space or extension became the fundamental reality, motion the point of all departure, and mathematics the language of its revelation. Descartes had made of nature, as J. H. Randall has put it, "a machine and nothing but a machine; purposes and spiritual significance had alike been banished. . . . Intoxicated by his vision and his success, he boasted, 'Give me extension and motion, and I will construct the universe.' "[12]

"Left Cartesianism": The Geometry of Politics

The modern dogma of scientism may be said to have sprung from the brow of Descartes almost, if not quite, full-panoplied. Only one thing in the world—the human mind—still escaped mechanical reduction and mathematical abstraction. But not for long. The original objection soon to be made against Descartes by the philosophers of England and the *philosophes* of France was not that he was too mechanical, but that he was not mechanical enough. The first and most formidable of these critics was Thomas Hobbes, who had been personally assured by Galileo himself of the correctness of his suspicion that "the sole and adequate explanation of the universe is to be found in terms of body and motion."[13] On a visit to the continent Hobbes familiarized himself with Descartes's *Meditations*, and promptly sat down to compose a vigorous set of "Objections" expressly for the author's benefit—in which he declared that mind and thought no less than other human activity were reducible to the motions of an animal organism: "If this be so, reasoning will depend on names, names on the imagination, and imagination, perchance, as I think, on the motion of the corporeal organs. Thus mind will be nothing but the motions in certain parts of an organic body."[14]

There was clearly no place in such a universe for dualism; no exceptions were to be allowed to the cosmic science of geometrical mechanics, whose truth had been conveyed to Hobbes by the successive revelations of Euclid and Galileo. With Hobbes, indeed, the mechanical philosophy came fully of age; and its major ramifications over the next three centuries are nearly all foreshadowed in his works. His theory of knowledge, in its tough-minded rejection of metaphysics and its insistence upon semantic precision, anticipates present-day logical positivism;[15] his psychology contains the mechanistic outlines of behaviorism; and his political philosophy presents a systematic portrait of that totally rationalized new order toward which the vision of modern behavioral scientists has turned no less irresistibly in the unending quest for certainty, predictability and control over the anarchic realm of politics and human affairs.[16]

After Hobbes the revisionist critique of Descartes from the "left" grew steadily in force until it reached its zenith in the authoritative strictures of Voltaire and d'Alembert against the exemption of *res cogitans*, the realm of thought, from objective measurement and mechanistic explanation. How much more congenial, for Voltaire, was the empirical Englishman Locke who had effectively "reduced metaphysics to what it ought to be in fact, the experimental physics of the soul,"[17] and so had driven the ghost from the machine. But long before the advent of the *philosophes*, "left Cartesianism" (as Lewis Feuer has aptly termed it) had emerged in full strength in seventeenth-century Holland—most significantly of all, in the geometrical ethics and politics of Spinoza.

"Spinoza begins," said Leibniz, "where Descartes ended, in Naturalism."[18] And where Spinoza ended, in his own turn, was with the vision of a determinate world order in which the reign of objective reason would usher in the millennium of universal happiness. Spinoza's scientific determinism was pure and total: nothing escaped its embrace, neither the soul of man nor God Himself. It comprehended a universe of relentless necessity; but it was a necessity emptied of purpose, a chain of effects without final causes. This achievement was, as Randall has seen, the first great revolution which Spinoza's science contemplated: the second lay in "placing man and his life at the very core of the great machine."[19] In what reads strangely like an anticipatory rebuttal of the quantum physicists of three centuries later, Spinoza fulminated against the advocates of free will.[20] "Most writers on the emotions and on human conduct," he declared, "seem to be treating rather of matters outside nature than of natural phenomena following nature's general laws. They appear to conceive man to be situated in nature as a kingdom within a kingdom; for *they believe he disturbs rather than follows nature's order.*" Not so Spinoza: "I shall consider human actions and desires in exactly the same manner as though I were concerned with lines, planes, and solids."[21]

Free will, then, was only an illusion of the human mind, a fiction exposed and repudiated by the new science. But that science also contained the means of attaining true freedom: namely, the posture of neutral objectivity and detachment. Freedom was, in other words, the recognition of necessity. Like the later determinists of scientific socialism, Spinoza preached the gospel of dispassionate objectivity with a subjective passion that has left him equally open to interpretation in the contradictory terms of science and mysticism.[22] But the paradox is scarcely more prominent in Spinoza than in the host of others before and since for whom the faith of scientism has become an evangelical gospel of salvation—promising, in its characteristic expressions, escape from the irrational freedom of the human will and the intolerable ambiguities of unscientific politics.[23]

Nor was Spinoza, for all his idiosyncracy, unique in his outlook among the Dutch intellectuals and statesmen (or even the merchants and mechanics) of his day. Professor Feuer has shown the extent to which, in its social as well as in its scientific aspects, Spinoza's philosophy was indebted to a leading Dutch Republican of the period, John de Witt. It was de Witt who first demonstrated the practical value of the geometrical method in the study of man and society.[24] The sociological imagination of de Witt was particularly displayed in an elaborate scheme which portrayed human societies as vast geometrical constructions, operating strictly in accordance with immutable natural laws, "to which the free will of each individual, after more or less variation, always ended by obeying."[25] Himself an eminent scholar and mathematician, it seemed axiomatic to de Witt that men of similar light and learning should guide the wills of ordinary individuals to this voluntary acquiescence in the general will. In fact, he invented an applied science of "social mathematics" according to which a body of social geometers would chart the future orbits of societies with all the precision of astronomers; and with social prediction would come, of course, political control.

The liberal republicanism of de Witt and Spinoza, like that of the rationalists of the next century, was never quite comfortable with the democratic prospect of universal enfranchisement and majority rule, but placed its faith instead in the enlightened administration of an aristocracy of talent drawn from the rising middle class and reflecting its already articulate ideology. "The god of Spinoza," remarks Feuer, "mathematical, scientific, had much that was congenial to the scientifically-minded merchants and artisans in the seventeenth century."[26] Most congenial of all, it was a god of mechanisms and calculations, free from the traces of scholasticism or mystery, whose Book was open to all (or at least to all who knew arithmetic) to read. This was an age of practical men and practical methods, in deliberate retreat from all that was subjective and intangible, or passionate and personal. At the midpoint of the seventeenth century the historian of the Royal Society in London could report the ambition of its scientists to be "to return back to the primitive purity and shortness, when men deliver'd so many things almost in an equal number of words." They had "a close, naked, natural way of speaking," and had succeeded in "bringing all things as near the Mathematical plainness as they can, and preferring the language of artizans, Countrymen, and Merchants, before that of Wits or Scholars." Philosophy itself, he concluded, would "attain to Perfection, when either the Mechanic Labourers shall have Philosophical Heads or the Philosophers shall have Mechanical Hands."[27]

The Quest for Certainty in the Heavenly City

Historians have not often given sufficient attention, as Lewis Mumford has pointed out, to the effect of the developing technics of the early modern period upon the concepts of political rule.[28] Well before the seventeenth century the methodical use of numbers, promoted by the Arabic system of notation, had begun to make its impact upon the human economy; while the introduction of town clocks as early as the thirteenth and fourteenth centuries came to symbolize a "methodical expenditure of hours" to match the "methodical accountancy of money." The emphasis of the merchant upon number and measure, and hence upon objectivity and impersonality, had remained a subordinate (if not subversive) theme in the universal order enforced by the Church; but by the seventeenth century little was left of that order and the respective roles of Church and tradesman were more nearly reversed—nowhere more markedly, indeed, than in the tolerant merchant-state of Holland. "After the Restoration," in the words of Tawney, "we are in a new world of economic, as well as of political, thought."[29] The claim of the Church to fix the rules of conduct in economic and political affairs was now all but ignored. The new and governing norms were those of modern capitalism in alliance with modern science. The moral rules of the Church gave way to "the new science of Political Arithmetic," which knew no morality beyond the letter of the law. Inspired by the contemporary progress of mathematics and physics, the new science handled economic phenomena exactly "as a scientist, applying a new calculus to impersonal economic forces."[30]

But it was not only economic forces which were treated thus impersonally by the objective calculus of Political Arithmetic. The new science promised the transformation of man himself, along with all of life, into the measurable and manipulable working parts of the great machine. If this had not been precisely the dream of Descartes, it was to become the all-absorbing vision of the post-Cartesians, brought to its fullest expression by Pierre Simon de Laplace—the pansophic vision of perfect prediction, exact measurement and absolute certainty. A superhuman Intelligence, proclaimed Laplace at the end of the eighteenth century, one which could know at a given moment all the forces by which nature is animated and the respective positions of the entities which compose it, "would embrace in the same formula the movements of the largest bodies in the universe and those of the lightest atom: nothing would be uncertain for it, and the future, like the past, would be present to its eyes."[31]

"Nothing would be uncertain for it." All phenomena were to be subsumed within the giant mechanism. What could not be stretched or shrunk to fit its procrustean framework was simply not "phenomena" at all, nor even noumena: it was only superstition, a kind of anamorphosis or fallacy of vision which the scientific lens would soon correct, and which meanwhile might better be disregarded.[32] In short, all that mattered was matter. All objects and fields of study—animal, vegetable or mineral—were equally and fully explainable by reduction to the impenetrable atoms which composed them and the physical forces which moved them. The ambition of the scientific true believers in the eighteenth century (as J. Bronowski has phrased it) was no less than "to impose a mathematical finality on history and biology and geology and mining and spinning."[33] The movement which had begun in the laboratories and the learned societies soon spread to all realms of intellectual life and became an urgent concern of civilization itself. It was d'Alembert, philosopher and mathematician, who most dramatically conveyed both the intensity and the universality of the scientistic spirit:

> Natural science from day to day accumulates new riches. . . . The true system of the world has been recognized. . . . In short, from the earth to Saturn, from the history of the heavens to that of insects, natural philosophy has been revolutionized; and nearly all other fields of knowledge have assumed new forms. . . . Spreading throughout nature in all directions, this fermentation has swept with a sort of violence everything before it which stood in its way, like a river which has burst its dams. . . . Thus, from the principles of the secular sciences to the foundations of religious revelation, from metaphysics to matters of taste, from music to morals, from the scholastic disputes of theologians to matters of commerce, from natural law to the arbitrary laws of nations . . . everything has been discussed, analyzed, or at least mentioned.[34]

Surely, as Carl Becker was to conclude, "this new philosophy ravished the eighteenth century."[35] Every man could understand, or thought that he could, the self-regulating mechanism described by Newton—or, better yet, by his countless popularizers.[36] Philosophers and humanists, artists and litterateurs, alike were drawn to become the students of natural science and frequently its expositors as well. Thus Voltaire took up the interpretation of the master text in his *English Letters* and *Elements of the Newtonian Philosophy*; Rousseau is said to have composed a tract on the laws of chemistry; Diderot wrote at length on the elements of physiology, and Montesquieu's early work was involved with physical and physiological problems.[37] Nor was this merely an eclectic display of intellectual competence in various and distinct fields of learning. In its characteristic expression it was the narrowest scientism: the systematic reduction of all subjects and fields of knowledge to the dimensions and categories of natural science. Philosophy tended to become "natural philosophy," biology virtually a branch of mechanics, and psychology the anatomy of the human machine.

The guiding metaphor of the new cosmology was given its most vivid and revealing expression by Fontenelle, in his *Conversations on the Plurality of Worlds*. The events of nature, he observed, may be likened to performances on a gigantic stage of which man is a spectator, alert to the panorama but indifferent to the technicalities of its production. But suppose there to be a "mechanic" in the audience; would he not yearn to get behind the scenes, to penetrate the mechanism of the stage production? The philosopher of the Age of Reason was just such a mechanic; his interest was not simply in the passing parade but in the clockworks which make it run. "I esteem the universe all the more," said Fontenelle, "since I have known that it is like a watch. It is surprising that nature, admirable as it is, is based on such simple things."[38]

The analogy of Fontenelle illuminates two of the essential attributes of the Newtonian world view: the image of the great machine itself, and the expulsion of man from the center of the stage—from an active part in the drama—to a seat in the audience and the passive role of spectator. The full meaning of this tableau was that man had disappeared from the world as *subject* in order to reappear as *object*. Mind itself was dissolved into particles in motion by the neutralizing solvent of the new physics. If this denouement had been anticipated by the mental geometries of Hobbes and Spinoza, it was in the century of enlightenment, with the appearance of such thoroughgoing reductions as those of Holbach and La Mettrie, that humanity was not merely subordinated to but absorbed within the giant machine. Man, according to Holbach, is entirely the work of Nature. "He exists in Nature. He is submitted to her laws. He cannot deliver himself from them."[39] All the processes of nature, moral and political no less than physical, were reducible to matter and motion and completely accounted for in mechanical terms: "The universe, that vast assemblage of everything that exists, presents only matter and motion: the whole offers to our contemplation nothing but an immense, an uninterrupted succession of causes and effects."[40]

"Let us conclude boldly then," added La Mettrie, "that man is a machine, and that there is only one substance, differently modified, in the whole world. What will all the weak reeds of divinity, metaphysic, and nonsense of the schools, avail against this firm and solid oak?"[41] The human organism was to be regarded as an automatic clock, so skillfully and "artificially" contrived that if the wheel governing the second hand should break down, the minute hand would carry on within its own mechanical orbit. A knowledge of the clockwork was all that was necessary to comprehend the full range of human behavior. "If we assume the least principle of motion, then animated bodies will have all they require in order to move, to feel, to think, to repent, and in short to conduct themselves physically and, which is dependent on this, morally."[42]

The consequence of such analogical reasoning, for those straightforward materialists, was obvious and inescapable: the universal pattern of rigid determinism, of mechanistic causation, applied to human nature as plainly as to physical nature. Man was the hapless creature, and human behavior the pawn, of forces set in motion at the beginning of time; the concept of freedom was no more than self-deception, and that of purpose only a superstition. "If man believes himself free, he is merely exhibiting a dangerous delusion and an intellectual weakness. It is the structure of the atoms that forms him, and their motion propels him forward; conditions not dependent on him determine his nature and direct his fate."[43]

On nearly every plane of scholarship, during the century after Newton, the evidence accumulated of man's subjugation to the laws of physics and chemistry. In biology the anatomical studies of Vesalius, and subsequently Harvey's discovery of the circulation of blood, had furnished a powerful stimulus to "the comparison between living organisms and machines working according to the laws of mechanics."[44] The psychologists in particular embraced the mechanistic viewpoint and joined in the quest for measurable certainty. Hartley, the founder of the associationist school and possibly the first Englishman to employ the title of "psychology" for his subject matter, sought to inaugurate a science of behavior which would be purely and simply an extension of physics; in his *Observations on Man, his Frame, his Duty, and his Expectations,* he openly imported Newton's method and language into psychology.[45] Earlier in the century a deeper intelligence had produced the *Treatise on Human Nature* expressly as "an attempt to introduce the Experimental Method of Reasoning into Moral Subjects." In this, as in his later *Inquiry into the Human Understanding*, David Hume hoped to duplicate in the realm of mind what Newton had accomplished for the realm of nature; in the principle of association he discerned "a kind of attraction, which in the mental world will be found to have as extraordinary effects as in the natural, and to show itself in as many and as various forms."[46] In psychology no less than other sciences, as Werner Heisenberg has observed of this period, everyone labored to apply the concepts of classical physics, and most of all that of causality. The life of the mind, like all of life, was to be explained as "a physical and chemical process, governed by natural laws, completely determined by causality."[47]

By the close of the eighteenth century Newton's method, the method of causes and mechanisms, had become standard procedure throughout the respective sciences of nature, of life and of man. Possessed of the key to universal knowledge and predictive certainty, Laplace could casually dismiss Napoleon's question as to the place of God in his System of the Worlds: "I have no need of that hypothesis." Surely, as Becker has concluded, the disciples of the Newtonian philosophy had not ceased to worship: "They had only given another form and a new name to the object of worship; having denatured God, they deified nature."[48] There was one further step to take, if only a short one; having dehumanized man, they erected mechanical models, imbued with a more objective rationality, to replace him.[49]

Social Scientism in the Century of Progress

In the course of the nineteenth century the systematic projection into the "humane studies" of the spirit and method of Newtonian physics was carried to its extreme in nearly every direction. The two fundamental postulates of the scientific mech-

anist—those of neutral objectivity and analytic reductionism—
came to be reflected, with varying degrees of accuracy and distor-
tion, in many of the most influential social theories of the period.
At the very outset of the century, Saint-Simon in his first published
brochure proclaimed the founding of the Religion of Newton. In
the Newtonian law he had made out the single principle governing
the entire phenomenal world: ". . . universal gravity is the sole
cause of all physical and moral phenomena. . . ."[50] All of human life
and activity, of society and politics, were reducible to the ultimate
source of cosmic gravitation (or, in the parallel system of Fourier,
to that of "passionate attraction"). In the development of his
scientific religion, as Frank E. Manuel has persuasively shown,
Saint-Simon was only following the inveterate tendency of his age
to identify a single principle underlying all reality—from which,
with perfect logic and accuracy, all the special sciences including
those of man and society might be successively unraveled.[51]

It was to be the specific task of Saint-Simon's "Supreme
Council of Newton" (assisted by innumerable local Councils
housed in a series of Newtonian temples) to educate the general
public to an understanding of the universal law. All this had been
revealed to the visionary by God Himself, who also confided the
information that He had placed Newton at His side and entrusted
him with the guidance not only of humanity but of all inhabitants
of the universe. Nor was there any place on the Supreme Council
for those who sought to cling to prescientific thoughtways; the
philosophers were driven like money-changers from the Temple of
Newton. "It is necessary that the physiologists chase from their
company the philosophers, moralists and metaphysicians just as the
astronomers and the chemists have chased out the alchemists."[52] In
fact, the scientific councilmen would rule with an iron hand and a
puritanical devotion: "All men will work; they will regard themselves as laborers attached to one workshop whose efforts will be
directed to guide human intelligence according to my divine foresight. . . . Anybody who does not obey the orders will be treated
by the others as a quadruped."[53]

In its subsequent revisions, the new world of Saint-Simon
took still more definite and authoritarian shape. The science of

sciences assumed the title of "Physicism" (later "Physicalism"), and became the ultimate stage of development both of science and religion. Now under the guidance of the Emperor Napoleon, "the scientific chief of humanity as he is its political chief," the religion of science was consciously aimed at *the golden age that is not behind us but in front of us and that will be realized by the perfection of the social order.*[54] This phrase was to become a maxim of the Saint-Simonians of the *École polytechnique*, and to serve as inspiration for that all-embracing science of society (first known as "Social Physics") to be undertaken by Saint-Simon's chief disciple, Auguste Comte. As we shall see in some detail later on, Comte's achievement was to bring system and plausibility to the hallucinatory visions of his master. "We shall find," he wrote, "that there is no chance of order and agreement but in subjecting social phenomena, like all others, to invariable natural laws, which shall, as a whole, prescribe for each period, with entire certainty, the limits and character of political action: in other words, introducing into the study of social phenomena the same positive spirit which has regenerated every other branch of human speculation."[55]

The political significance of the Saint-Simonians (as such present-day students as Salomon and Hayek have perceived) was that their zealous effort to apply the Newtonian principle to the study of society was never simply contemplative but consciously manipulative in design.[56] Their vision was of a society wholly made over in the image of the new mechanics—technically rationalized in every detail, predictable in every activity, and hence brought under total scientific management. The religion of science was a faith in the existence of an objective Reason, impersonal and mechanical, harmonious and determinate, existing entirely apart from individual men and indifferent to their purposes. But where the Cartesians had found its location in the Great Machine, and the *philosophes* had seen it to reside in Nature, the Saint-Simonians and their sociological successors placed its residence in "Society"—not, to be sure, in the confused and imperfect societies of historical reality, but in a "New Atlantis" administered by a council of social physicists and presided over in spirit by the demon of Laplace.[57]

While Saint-Simon and his followers in France were constructing the positive science of society, Jeremy Bentham and his Utilitarian disciples across the channel were no less ambitiously engaged in an imaginative reconstruction of political and social institutions on very much the same foundation. "What is known as Utilitarianism, or Philosophical Radicalism," the foremost historian of the movement has concluded, "can be defined as nothing but an attempt to apply the principles of Newton to the affairs of politics and of morals."[58] The first step, for James Mill as for Bentham, was to devise a mechanistic explanation for mental events (by means of associationism); next to do the same for social events; and finally to erect a scaffolding of moral and legal theory which would present not only explanation but vindication of the program on strictly scientific grounds. In this Utilitarian system human beings were frankly viewed as social atoms, each of the same measurable weight and specific gravity, all alike motivated by the mechanical "springs of human action": that is, the pursuit of pleasure and the avoidance of pain. Into this aboriginal stimulus-response psychology was next introduced a scientific means of conditioning human responses: the Felicific Calculus, by means of which a benevolent legislator might "maximize happiness by a sort of arithmetics of pleasure."[59] Indeed, once the conception of atomistic individualism (formulated by the physiocrats and systematized by Adam Smith) had taken root, the new science of Political Arithmetic gained almost unchallenged acceptance in economic and social thought.[60]

One event particularly, in the evolution of classical economic theory, stands out as a watershed in what Karl Polanyi has termed the "great transformation" of nineteenth-century social thought from a humanistic to a naturalistic orientation. Although Adam Smith, as a good citizen of the "heavenly city," had sought to be as scientific as possible in his theorizing, the science which he brought to political economy was still essentially a *human* science, based upon the peculiar characteristics of man rather than upon those of external Nature. For Smith the realm of economic phenomena was still a part of "politics"—that is, of human contrivance and control. But, as had happened earlier with the Cartesians' theory of nature, Smith's conception of the market mechanism was not quite mechanical enough to suit the mood of the age—or, more concretely, to fit the demands of the marketers and the felt regularities of the marketplace. A more appropriate and congenial hypothesis appeared ten years after *The Wealth of Nations* in the form of Joseph Townsend's *Dissertation on the Poor Laws,* which presented an account of economic balance altogether independent of human effort and intervention—a balance arising wholly from natural forces and biological conditions.[61] In Townsend's pre-Darwinian parable of the island of Juan Hernandez—an animal kingdom inhabited by goats and dogs—there was no government or law, nor any agency of purpose, yet a perfect equilibrium was perpetually maintained: the natural balance of the struggle for survival.

Townsend's dissertation was, of course, not a brief for more government but for less. In positing the existence of a natural mechanism automatically regulating economic society, Townsend laid the groundwork for the emancipation of economics from government and politics—and with economic theory went much of social thought.[62] The simple thesis which Townsend advanced —and which the classical economists who followed him were to confirm and elaborate for the next hundred years—was that economic society was founded on the brute realities and processes of nature, whose laws were definitely *not* human laws. The acceptance of this thesis, as Polanyi has shown, ushered in a transformation not only of economic thought but of the intellectual consciousness of the nineteenth century. "From this time onward naturalism haunted the science of man. . . ."[63]

The corollary of the new portrait of economic society as an automatic and autonomous mechanism was, of course, the now-classic image of Economic Man. The widespread application of Bentham's Felicific Calculus to human behavior—meticulously reckoning all that was reducible to quantity and discarding all that was not—led directly to the inference which was to become embodied in the conventional wisdom of classical economics and straightforwardly announced by Francis Edgeworth in his *Mathe-*

matical Psychics: namely, that man is a "pleasure-machine."[64] Stimulated by the instinctual urges of pleasure-and-pain (or profit-and-loss), Economic Man became an objectively calculable unit of economic society—no less determinate and predictable in his conduct, because no more consciously self-directed, than any other component of the impersonal market.[65]

Darwinism: Social and Anti-Social

During the course of the nineteenth century, two opposed and seemingly irreconcilable viewpoints persistently competed for acceptance in the various arenas of social thought. On the one hand, atomistic individualism portrayed human society as an aggregation of disparate individuals, virtually as self-contained and "window-less" as Leibniz's monads—and characterized especially by a pro-nounced aversion to political control. On the other hand, theories of group or societal primacy thoroughly subordinated the individ-ual to the purposes of some larger unity, generally culminating in the glorification of nation, society or state. It has been the custom among historians to stress the contrasts and polarities separating these two conceptual frameworks, both in theory and practice; but the neatness of the contrasts has always been marred by an incorrigible tendency of the "poles" toward convergence. Thus Utilitarianism, for example, embodied at the same time a theory of laissez faire and a doctrine of parental government;[66] while, some-what earlier, Rousseau could plausibly appear as both the apostle of liberty and the prophet of a coercive collectivism.

Both sides of the debate, after the seventeenth century, felt equally entitled to claim the authority of classical physics for their premises; and in the nineteenth century both of them extrapolated eagerly from the biological evidence of Darwin. It does not follow from this, however, that one or the other viewpoint was con-sciously falsifying its credentials or acting cynically in its claims. Both contending philosophies were sincere in their belief that they were faithfully applying the principles of Newton and Darwin;

and in one significant particular, at least, they were in complete accord. Even in their most extreme expressions—those of atomistic individualism and organic collectivism—both perspectives were fundamentally at one in their assumptive *image of man*. Although they differed widely in their eschatologies—in their anticipations of human destiny—they were agreed in their underlying psychology: specifically, in the belief that the behavior of the individual human being is the product of circumstances beyond his control and is therefore at bottom involuntary, irresponsible, and mechanical.[67]

Perhaps the most striking evidence of this shared framework of assumption is to be seen in the divergent careers of the idea of "social Darwinism" under the respective guidance of individualists and group-theorists. Much the better-known of the two, of course, is the individualistic movement associated with such spokesmen as Spencer, Huxley and Sumner, which became the systematic apologetics of liberal capitalism and conservative politics. It is probable, however, that the attraction of this evolutionary scientism lay scarcely more in its economic implications than in its biological confirmation of the purposeless universe of Newtonian mechanics. The popular Darwinism of Spencer and Huxley, in Jacques Barzun's words, "captivated a generation of thinkers whose greatest desire was to get rid of vitalism, will, purpose or design as explanations of life, and to substitute for them an automatic material cause."[68] In short, still more convincingly than had been possible in the past, the theory of organic evolution subjugated man to nature and its mechanical laws.

This subordination was nowhere more earnestly carried out than in the prolific addresses and writings of Thomas Henry Huxley, who was prepared to define mankind simply as "conscious automata" and to maintain that "all vital action may . . . be said to be the result of the molecular forces of the protoplasm which displays it." It followed (as he announced in a famous lecture) that "the thoughts to which I am now giving utterance, and your thoughts regarding them, are the expression of molecular changes in that matter of life which is the source of our other vital phenomena." Huxley's central message was that of a scientific imperialism intent upon extending its hegemony to the farthest reaches of mind and spirit. The progress of science meant nothing more nor less than "the extension of the province of what we call *matter* and *causation*, and concomitant banishment from all regions of human thought of what we call *spirit* and *spontaneity*."[69]

But it was in the hands of Herbert Spencer that the evolutionary hypothesis found its most ambitious and sweeping formulation: an effort to bring together, within one Synthetic Philosophy, all the knowledge of the natural sciences—and to focus it like a searchlight upon the nature and destiny of man. All that Spencer awaited was a unifying formula within whose rubric the findings of the various sciences might be brought into harmony. "The question is, not how any factor, Matter or Motion or Force, behaves by itself. . . . Only when we can formulate the total process, have we gained that knowledge of it which Philosophy aspires to." The formula which emerged was that of the evolution of species from the simple to the complex, faithfully adapted to human society and grimly redefined in the physicalistic language of matter and motion: "Evolution is an integration of matter and concomitant dissipation of motion; during which the matter passes from a relatively indefinite, incoherent homogeneity to a relatively definite coherent heterogeneity; and during which the retained motion undergoes a parallel transformation."[70]

This progression from homogeneity to heterogeneity was for Spencer as inevitable as it was natural; no agency, human or other, could alter or improve it. It worked out its inexorable career alike in the evolution of species, in the development of mind, and in the progress of societies. But it could not continue forever onward and upward; it must eventually culminate, in its social phase, in the attainment of "equilibration"—that is, of a stable and harmonious state bringing with it "the establishment of a greatest perfection and the most complete happiness."[71] Almost a decade before the appearance of Darwin's theory, Spencer had set down his vision of the perfect society as a necessary outcome of the biological adaptation of man to nature:

The ultimate development of the ideal man is logically certain —as certain as any conclusion in which we place the most implicit faith; for instance that all men will die. Progress, therefore, is not an accident, but a necessity. Instead of civilization being artificial, it is a part of nature; all of a piece with the development of the embryo or the unfolding of a flower.[72]

It is plain from this that, whatever their disagreement on the role of government in the future utopia, Spencer and Comte were similarly possessed by the idea of progress as the certain and predictable fulfillment of natural scientific laws. Despite his self-conscious individualism, Spencer's vision of the millennium (perhaps reflecting his early training as a civil engineer) was as technological in spirit as that of the Saint-Simonians—no less intolerant of ambiguity, and, in a word, no less authoritarian. If it was not exactly a welfare state, this was because no such artificial ministrations were necessary; the fittest would continue to survive until all who survived were perfectly fit.

Among the "individualistic" Darwinians, it was the American William Graham Sumner who was most forthright in spelling out the moral which all alike accepted: that the individual human being was the passive agent of natural forces working through him, which knew nothing of his purposes and were oblivious to his will. To be sure, the person played a key role in the drama of evolution, but his was a mechanical performance in a dumb show whose plot remained a riddle and whose outcome he could not hope to affect. "We are convinced that this way of looking at things frees our treatment from a current tendency, which we regard as confusing and unproductive, to refer societal results to conscious, reasoned, and purposeful action on the part of the individual."[73]

Sumner's thoroughgoing pessimism regarding "the absurd attempt to make the world over" was balanced by a deep faith in the rightness of things as they were, along with a comforting counsel of relaxation in the cradle of the evolutionary process. All efforts to divert the tide or harness it to human purposes were futile; man was the creature, not to say the prisoner, of his natural history, from whose unsentimental embrace there could be no escape. "The great stream of time and earthly things will sweep on just the same in spite of us. . . . Every one of us is a child of his age and cannot get out of it. He is in the stream and is swept along with it. All his science and philosophy come to him out of it. Therefore the tide will not be changed by us. It will swallow up both us and our experiments."[74]

Among these transient experiments was that of political democracy, the "pet superstition of the age," which on its humanitarian side—in its emphasis on equality, opportunity and the rights of man—Sumner heartily detested and excoriated. "There can be no rights against Nature," he asserted, "except to get out of her whatever we can, which is only the fact of the struggle for existence stated over again."[75] What men mistook for natural rights and moral values were only the "rules of the game of social competition which are current now and here." Political ideals were unscientific fantasies, devised to pacify the restless and avoid the tackling of practical problems.[76] The desire for equality was a "superstitious yearning"; there could be no place whatever for sentiment in the conduct of human affairs; nothing but might had ever made right, nor ever would.[77] All the cherished notions associated with the dignity of men were only passing fancies, the ephemera of the folkways; and Sumner looked forward with barely disguised impatience to their passing.[78]

The social-Darwinist movement of which Huxley, Spencer and Sumner were the most prominent spokesmen was, at bottom, less a celebration of "individualism" than of impersonal mechanism. The human individual was no more the center of its philosophy and the source of its values than the atom was the moving force of the Newtonian universe; the instrumental factor, the only significant reality, was the evolutionary mechanism of nature. Man was merely a somewhat conscious automaton whose very consciousness was itself an efflorescence of protoplasm and a captive of the folkways.

It is notable that these were also the conclusions reached by the *anti*-individualist wing of social Darwinism, which emphasized the "group" (variously defined as class, race, society or nation) as the agent of evolutionary struggle and progress. It was in Germany and Austria that the theory of Darwin was turned into

Darwinismus, a vast extrapolation of the biological hypothesis to all realms of culture—for which Chauncey Wright coined the term "German Darwinism" and observed that the same thing was happening to Darwin at the hands of German metaphysicians as had happened to Newton at the hands of his French philosophical disciples.[79] The most influential of the German Darwinists was probably Ernst Haeckel, who made use of natural selection as the basis for a sweeping mechanistic philosophy intended to unite organic and inorganic nature and to reduce the highest activities of mind and culture to material changes in the protoplasm.[80]

It remained, however, for the Austrian sociologist Ludwig Gumplowicz to discover in evolution the inspiration for a theory of group conflict and social determinism which was outspokenly contemptuous of the value of individual human life. For Gumplowicz all history was the history of group struggle, governed by no other interest than power and no other cause than blind natural law. "In truth," he wrote, "everywhere and from the very beginning the social world has moved, acted, fought and striven only by groups."[81] In the ceaseless struggle of the groups individual opinions play no part; "blind natural law controls the actions of savage hordes, of states and of societies." The behavior of the individual, in thought no less than action, was to be seen as an automatic response to outer stimuli predetermined according to fixed laws:

> *The great error of individualistic psychology is the supposition that man thinks.* . . . A chain of errors; for it is not man himself who thinks but his social community; the source of his thoughts is in the social medium in which he lives, the social atmosphere which he breathes, and he cannot think ought else than what the influences of his social environment concentrating upon his brain necessitate. . . .
> The individual simply plays the part of the prism which receives the rays, dissolves them according to fixed laws and lets them pass out again in a predetermined direction and with a predetermined color.[82]

Since it is nature that rules all of life and history, human acts are "reasonable" only when they conform to natural tendencies, and unreasonable when they mistake or oppose them: "There can be but one principle of human rationality and of human morals and ethics: to be governed by the import and tendency of nature's sway. Hence knowledge of nature, natural science in its full scope, embracing every department of human life, is the only and the necessary basis of the science of morals and ethics." Sociology in particular, as the natural science of society, thus became a moral science preaching "man's renunciatory subordination to the laws of nature which rule history."[83]

But Gumplowicz's sociology also preached a more significant submission. Observation of the social world revealed to him "a necessity immanent in the condition of things" which moved inexorably toward fulfillment.[84] This necessity was visible in the movements of society, and more particularly in the actions of the state. It was the state which conferred all rights and embodied all justice; there were no rights anterior or superior to its mandate. "The premises of 'inalienable human rights,' " he maintained, "rest upon the most unreasonable self-deification of man and overestimation of the value of human life. . . ."[85] This fancied freedom and equality was simply incompatible with the state and indeed was its direct negation. The only real choice open to men was between submission to "the state with its necessary servitude and inequality, and—anarchy."[86]

The significance of Gumplowicz's unrelenting fatalism, for our present purposes, is in the remarkable consistency with which he pressed to its logical conclusion the human inference from evolution regarded as a mechanical process. For him, as for the mainstream of philosophy and social thought in the second half of the century, Darwin's discovery rounded out the cosmological design first charted by Newton; it furnished the basis for the final reduction of the organic and human world to the physical laws governing the inorganic universe. As Ernst Cassirer has seen in summarizing the philosophical impact of Darwin, "It is the same iron ring of necessity that encloses both our physical and our cultural life. In his feelings, his inclinations, his ideas, his thoughts, and in his production of works of art, man never breaks out of this magic circle."[87]

The convergent influences of Darwin and Newton were meanwhile to be found elsewhere in the social thought of the century—most notoriously of all, perhaps, in the scientific socialism of Marx and Engels. Marx prefaced his volumes on *Capital* with the words: "It is the ultimate aim of this work to lay bare the . . . law of motion of modern society."[88] Engels was equally explicit in his repeated assertion that "Marxian dialectics is nothing more than the science of the general laws of motion and development of Nature, human society and thought."[89] All life and behavior, in the scheme of Marxist materialism, were ultimately to be explained in the dialectical terms of matter in motion. Thought itself was a product of this natural process, and man was inescapably subject to "the general laws of motion and development" which govern the material universe; in their necessity lay his freedom.[90]

On still another plane of scholarship, the effects of social scientism were on display during the century in the aggressive career of "scientific" history, under the influence and example principally of Leopold von Ranke. In this rigorous school the historian was taught to reject all subjective considerations of judgment and value in favor of monastic devotion to the postulates of detachment and determinism.[91] Under the banner of exact history (*"wie es eigentlich gewesen ist"*), the disciples of Ranke dominated historiography on the continent, in Britain, and finally in America until well after the turn of the century—when it was still possible for J. B. Bury to announce that "history is a science, no less and no more. When Ranke's dictum is taken to heart, there will no longer be divers schools of history."[92]

The visionary goal of scientific historians of the nineteenth century was retrospectively defined toward the end of the cycle by one of the most talented and ambitious of their number, Henry Adams. Four out of five serious students of history still living, he wrote,

> have felt at some time in the course of their work that they stood on the brink of a great generalization that would reduce all history under a law as clear as the laws which govern the material world. . . . The law was certainly there, and as certainly was in

places actually visible, to be touched and handled, as though it were a law of chemistry or physics. No teacher with a spark of imagination or with an idea of scientific method can have helped dreaming of the immortality that would be achieved by the man who should successfully apply Darwin's method to the facts of human history.[93]

One final illustration deserves to be mentioned which demonstrates, perhaps more strikingly than any other, the extent to which the objective and mechanical spirit of natural science captivated the consciousness of the nineteenth century. In the year 1895 Sigmund Freud completed the preparation of a manuscript entitled "Project for a Scientific Psychology," whose opening words were unequivocal: "The intention of this project is to furnish us with a psychology which shall be a natural science: Its aim, that is, is to represent psychical processes as quantitatively determined states of specifiable material particles and so to make them plain and void of contradictions."[94] Freud's projected "natural science" of psychology was never published, and in fact was discarded almost as soon as it had been set down. But that it should have been conceived at all is peculiarly suggestive of the degree to which even this great protagonist of the independent reality of the human mind was, nevertheless, a child of his time.[95]

With these successive and, for the most part, triumphant invasions of the main currents of social thought during the nineteenth century, the world view of classical physics—the image of the Great Machine—was extended to its logical, psychological and sociological limits. Nearly everywhere, by the end of the century, the dominant impulse seemed to be to force the objects of human inquiry and concern beneath the microscope of mechanistic analysis, to reduce their content to the smallest measurable denominator or the single irreducible cause—without, at the same time, contaminating the observation with "subjective" considerations. The ideal of the social scientist in the Age of Progress was still very much as Fontenelle had characterized it: merely to be a spectator at the grand performance of Nature—but a spectator with the mentality of a mechanic.

The consequences of all this for the contemporary enterprise of social science—for those interrelated studies once regarded as "humane" but now more widely designated as "behavioral"—form the subject matter of the chapters to follow. Something of their general theme may perhaps be anticipated, and that of the present chapter restated, through this summation by a brilliant biologist of our own day, Ludwig von Bertalanffy, which stands as the conclusion of his *Problems of Life*:

> The evolution of science is not a movement in an intellectual vacuum; rather it is both an expression and a driving force of the historical process. We have seen how the mechanistic view projected through all fields of cultural activity. Its basic conceptions of strict causality, of the summative and random character of natural events, of the aloofness of the ultimate elements of reality, governed not only physical theory but also the . . . viewpoints of biology, the atomism of classical psychology, and the sociological *bellum omnium contra omnes*. The acceptance of living beings as machines, the domination of the modern world by technology, and the mechanization of mankind are but the extension and practical application of the mechanistic conception of physics.[96]

CHAPTER II

The Alienated Machine:
Psychology as the Science of Behavior

The hypothesis that man is not free is essential to the application of scientific method to the study of human behavior.
—B. F. Skinner

In short, the cry of the behaviorist is, "Give me the baby and my world to bring it up in and I'll make it crawl and walk; I'll make it climb and use its hands in constructing buildings of stone or wood; I'll make it a thief, a gunman, or a dope fiend. The possibility of shaping in any direction is almost endless."
—John B. Watson

I T IS A CURIOUS FACT that most attempts to account for the origins and development of the contemporary phenomenon known as behavioral science have paid little or no attention to its closest relative: psychological behaviorism. Where it is mentioned at all, behaviorist psychology tends to be hastily dismissed as a radical and abortive enterprise of an earlier day which has left no lasting deposit in the minds of contemporary psychologists.[1] But the demonstrable truth is, on the contrary, that the basic tenets of behaviorism have not only been kept alive in the laboratories of experimentalists but hold as secure and prominent a position as ever in their conceptual schemes. Moreover, it is hardly too much to say that the most militant leadership in the general movement of behavioral science has been furnished by the advocates of its thriving namesake in psychology.[2]

As recently as 1957, the members of the American Psychological Association presented an official citation which read: "To Dr. John B. Watson, whose work has been one of the vital determinants of the form and substance of modern psychology. He initiated a revolution in psychological thought, and his writings have been the point of departure for continuing lines of fruitful research."[3] A few years before, one of the most respected of contemporary behaviorists had insisted that, while some changes have been rung upon the original Watsonian system over the intervening generation, "many of the basic postulates of his formulation are to be found in the present-day varieties of behaviorism and, what is more important, probably, in the underlying working assumptions of the great majority of present-day American psychologists."[4] In light of these and similar accolades, it should be of more than historical interest to seek an understanding of the "basic postulates," as well as the assumptions and ambitions, of this acknowledged pioneer of the modern science of psychological behavior.

The Background: Association and Utility

Perhaps the first point which needs to be underlined is that—contrary to the official citation of Watson and to the manifestoes of his early partisans—behaviorism was from the outset less a revolution than a *counter*revolution in psychological thought. Although it emerged as a deliberate protest against the preoccupation of much nineteenth-century psychology with consciousness and introspection, behaviorism was essentially a rehabilitation of the very much older associationist tradition. As we have seen in Chapter 1, associationism in its classical form represented an attempt to translate the postulates and methods of Newtonian mechanics, as precisely as possible, into psychological terms. For Hobbes (who might be regarded as the original behavioral scientist), all mental operations were reducible simply to the "motions" of natural events; his psychology was "materialistic, mechanistic, and deterministic throughout"[5]—an uncompromising effort to encompass all aspects of human existence within the iron law of physical explanation.

In similar fashion Locke's atomistic empiricism, while less dogmatic and logically coherent, was instrumental in the shaping of the associationist tradition through its assumptive image of human nature as a passive agent, a *tabula rasa*, upon which the irresistible forces of outer experience work their will and have their way. Lockean reductionism, as Gordon W. Allport has reminded us, operated in at least three ways: what was *small and molecular* ("simple ideas") was more fundamental than what was large and molar ("complex ideas"); what was *external and visible* was somehow more important than what was not; and what was *earlier* in development was more basic than what came later.[6] The mind was treated by Locke and his fellow empiricists, to use the figure of Isaiah Berlin, "as if it were a box containing mental equivalents of the Newtonian particles." Those equivalents, the "simple ideas," were indivisible atomic entities originating somewhere in the external world, "dropping into the mind like so many grains of sand inside an hourglass."[7]

The associationist school which arose from these sources, and came to dominate British psychology for upwards of two centuries, was therefore securely grounded upon a substructure of mechanistic assumptions.[8] During the course of the eighteenth century, both Hume and Hartley strove to erect upon the principle of association an exact natural science of the mind which would be as "positive" and objective as the axioms of the new mechanics.[9] Hartley not only took over his method and terminology straight from Newton but combined his psychological theory with a physiological doctrine whose central idea was no less Newtonian—one in which "vibratiuncules" or "vibrations in miniature" were substituted for the Cartesian traces.[10] This psychophysical parallelism did not, to be sure, go quite so far as to assert the primacy (let alone the exclusive reality) of the physical functions accompanying mental processes—as Hobbes for example had done, and as the mechanistic school of French psychology was seeking to do in Hartley's time. (The physiologist Cabanis, for one, following the lead of La Mettrie's *L'Homme Machine*, declared unhesitatingly that all the aspirations and achievements of the human spirit are properly to be seen as the efflorescence of physical functions and the outcome of immutable laws of nature.)[11]

The full flowering of Hartley's physiological approach to psychology, however, awaited the advent of British philosophical radicalism at the outset of the nineteenth century. Throughout its vigorous career it was the inherent tendency of Utilitarian psychology, as its foremost historian has shown, "in some sense to materialize thought in order to find for the invisible and intangible psychological phenomenon, some palpable equivalent which can get hold of the methodical observations of the enquirer."[12] Erasmus Darwin, in his *Zoonomia*, anticipated the most ambitious thrusts of twentieth-century behaviorism by devising a theory of sensations and the association of "ideas" which abolished mind altogether and traced its apparent activity to biological motions.[13]

But it was Bentham's most dedicated disciple, James Mill, who carried the associationist scheme to its logical culmination. For Mill the human mind was not merely to be likened to a machine; it was in simple fact a machine, neither more nor less—a delicate mechanism whose clockwork operations were automatically triggered by physical forces from the outside (sensory stimuli), and kept going by forces no less physical on the inside.[14] It was this straightforward mechanical image of human nature which underlay the ultimate ambition of the philosophical radicals: to construct a science of legislation, and so of politics, resting squarely upon the foundation of the natural sciences. The very possibility of such an objective social science, of course, presupposed its accomplishment in the realm of psychology. Bentham, building upon Hartley's system in his *Table of the Springs of Action*, pointed the way; but it was the elder Mill who undertook systematically, in his *Analysis of the Phenomena of the Human Mind*, "to destroy the illusion of psychical activity . . . to reduce everything to constant and in some sort mechanical relations between elements which should be as simple as possible. . . ."[15]

Thus, in the forefront of the Utilitarian campaign to produce a natural science of man and society was the postulate of mechanistic reductionism: the analytical breakdown of all objects of investigation into their simplest observable parts or elements. The other, and no less central, postulate which we have also seen to characterize the world view of scientism—that is, the image of the scientist as an objective and disinterested spectator—was for the Benthamites rather more a tacit assumption than a systematic construct. It remained for a youthful French admirer of Bentham, Auguste Comte, to place deliberate emphasis upon this corollary aspect of scientism in the development of his own magisterial program to found a positive science of society—and a positive religion of Science.[16]

Comte: The "Tyranny of Progress"

Comte's famous "law of three stages" did not, as a matter of fact, allow any room at all within its organized hierarchy of sciences for "psychology" in its prevailing definition as the introspective study of mind. As the "facts" of any science were solely those gathered through external observation, so for scientific purposes the "inner life" was held to be simply inaccessible and irrelevant. "True observation," Comte declared, "must necessarily be external to the observer"; therefore "the famous internal observation is no more than a vain parody of it," which presents the "ridiculously contradictory situation of our intelligence contemplating itself during the habitual performance of its own activity."[17] Introspective psychology was to be regarded as nothing more than "the last transformation of theology," and thus could play no part in the proper study of mankind. But for Comte's positive scientist there remained two ways by which the phenomena of the mind might be rendered somewhat accessible: through study of the organs producing them ("phrenological psychology"), or through observation of "their more or less immediate and more or less durable results" (i.e., in behavior).[18]

The latter technique of positivist psychology (as F. A. Hayek among others has seen) is scarcely to be distinguished from what is known today as the behaviorist approach.[19] Although Comte, like later behaviorists, was far from consistent in his efforts to expel consciousness and purpose from positive science, he was at least consistent (until the last years of his life)[20] in seeking to derive all knowledge of man as an individual exclusively from biology and physiology. In this, as in other aspects of his "Social Physics," Comte was the heir of the Utilitarians and their predecessors of the associationist tradition; while in his vigorous insistence upon neutral detachment and "psychological distance" between observer and observed, he was the forerunner of subsequent positivist movements in the social sciences—movements which, while protesting their indifference to considerations of human value, have always done so on the basis of a rather fervent faith in the liberating value of their enterprise.[21]

However, the most significant (and unintentional) tendency of Comte's pre-behavioral science, together with that of the Saint-Simonians generally, was what Albert Salomon has called the "totalitarian potentiality"[22] implicit in its simultaneous celebration of society and nullification of the individual. The rejection of "psychology" (consciousness) was in effect a denial of rationality and of all distinctively human qualities; while the reduction of man to a merely biological object restricted him also to a passive and irrelevant role in the ongoing movement of social progress. A striking expression of this view was enunciated by Comte's contemporary, Maine de Biran:

> Not the human mind, not the individual understanding are the true subjects of the notions and verities of human existence. Society, however, gifted with a kind of collective mind, different from the individual's, is imbued with such knowledge. The individual, the human being is nothing; society alone exists. It is the soul of the moral world. It alone has reality, while individuals are only phenomena. . . . If this is true, all philosophy of the past was

wrong. One must recognize the failure of the science of the intellectual and moral man, one must admit the failure of psychology which has its basis in the primitive fact of conscience.[23]

The inevitable progress of society in future—specifically, an industrial society positively engineered on scientific lines—could receive nothing from the participation of individual men, beyond their recognition of the mechanical laws which ruled all events.[24] Indeed, declared Comte, "true liberty is nothing else than a rational submission to the preponderance of the laws of nature, in release from all arbitrary personal dictation."[25] For Comte no less than for Hegel, man acquired a degree of reason only through submission to the rational processes of society. As the source of reason was thus displaced from the individual to the collective, so liberal-democratic conceptions of personal freedom and responsibility gave way to an emphasis upon authority and social order—more exactly, upon the efficiency and rationalization of social engineering. In Salomon's words,

> The sociologists believed that a new world lay before them in which scientific planning, technical rationalization, and humanitarian education would be directed by anonymous social scientists who were subject to the laws of nature and of society, but not to the benighted authority of philosophers.[26]

It was a tragic vision. The goal of the positive scientists of society was nothing less than the universal enlightenment and emancipation of mankind. But the outcome was otherwise. Science, enlisted in the cause of reason, led in the end to the denial and abolishment of reason: "The logic and tyranny of progress gave to the world the progress of total tyranny."[27] The first comprehensive attempt to apply scientific method to the rationalization of human conduct—what might be termed the first systematic program of behavioral engineering—turned out to be, not a dispassionate and positive science of behavior, but a wholly passionate and negative campaign to make men *behave*.[28]

The Human Animal

One other direct precursor of behaviorism during the
nineteenth century is worthy of our notice: the field of animal
psychology which received its major impetus from the Darwinian
revolution in biology. For psychologists, the most pertinent effect
of the evolutionary hypothesis was its demonstration of the intimate continuity between the human and animal species, and consequently its vast encouragement of research upon rats, cats,
chickens and chimpanzees as surrogates for the human subject.
Moreover, insofar as animal psychologists could claim a closer
approximation than their introspective colleagues to the methods
of natural science (and hence a greater prestige in the scientific
hierarchy), their procedure became one devoutly to be wished by
the more backward study of human psychology.[29] At the hands
of the early Darwinians, to be sure, these researches seemed mainly
bent upon elevating the animals to the level of mankind, by ascrib-
ing to them all manner of complex mental achievements. Soon,
however, under the influence of such pioneers as Lloyd Morgan
and Thorndike, investigators subjected themselves to the law of
parsimony and discovered more elementary explanations for animal
behavior. Thereafter the direction of their efforts was reversed;
consistent with the parsimonious viewpoint, experimental psychologists appeared to be bent upon reducing mankind to the level of
the beasts.[30]

The development of animal psychology was, of course,
part of a pervasive movement of thought within the life sciences
during the latter half of the nineteenth century—a movement
spurred by the cumulative advance of exact methods in the study
of organic processes, and increasingly intolerant of any hint of
dualism or vitalism that might deter the complete absorption of
all living things, and of mind itself, within the orbit of natural
(i.e., physical) law. Borne forward on a wave of discoveries and
successes, this trend in the life sciences led inevitably, as Gardner
Murphy has recalled, to "a triumphant 'materialism,' or 'mecha-

nism,' in which the problems of mind were no longer to be tolerated as such, but were to be reduced to the form of general physical problems."[31] By the turn of the century the ground was well prepared for the emergence of a new psychology—a "psychology without a soul"—which would revitalize the mechanistic approach of the older associationism through the infusion of modern biological science, and so "systematically fulfill the promise of Hobbes and La Mettrie."[32]

The scientific psychology which arose with the new century to perform this evangelical role was that of behaviorism. Its path was finally cleared by a further auspicious development—the almost accidental discovery, by Russian physiologists working with animals, of the conditioned reflex.[33] Introduced by Pavlov in 1902, this concept came to be regarded by behaviorists as virtually sufficient for the total explanation—and, more significantly, for the prediction and control—of human behavior. There was, of course, little that was novel even at the turn of the century about the S-R formula—as early as the nineties Dewey had criticized the reflex-arc concept in an essay which has since become a classic[34] —but there was a powerful impression of novelty in the interpretation of these organic phenomena, and the sweeping claims made for them, by the behaviorists.

Watson: Behaviorism as Social Conditioning

Although Pavlov himself cautiously refrained from drawing psychological deductions from his animal researches,[35] his findings were seized upon by experimental psychologists, most of all in America, as the long-awaited key to a truly rigorous science of behavior. Particularly ambitious in drawing out these implications was John B. Watson, an experimentalist whose training was in the animal field.[36] Convinced that the objective physiological method could be applied without reservation to "the full range of human adjustments"—and seeking above all to gain for psychology the standing of a natural science—Watson did not hesitate to take the step of repudiating all reference to consciousness or purpose as irrelevant to the scientific appraisal of man.[37] For Watson "psychology, as the behaviorist views it, is a purely objective, experimental branch of natural science which needs introspection as little as do the sciences of chemistry and physics."[38] In short, if psychology were to become a natural science, it must (in Heidbreder's summation) "become materialistic, mechanistic, deterministic and objective."[39]

The preliminary step in gaining scientific status was to renounce the old psychology root and branch. The behaviorist began, wrote Watson, "by sweeping aside all mediaeval conceptions. He dropped from his scientific vocabulary all subjective terms such as sensation, perception, image, desire, purpose, and even thinking and emotion as they were subjectively defined."[40] Behaviorism as a concept came to stand for the deliberate limitation of scientific interest to the objectively observable. "Now what can we observe? We can observe behavior—what the organism says and does."[41] The scope of this observation did not, of course, include what the human organism *means* or *intends* by his speech and action; "meaning" in fact had nothing to do with subjective intentions—it was a property not of the behaver but only of his "behavior." In effect, the meaning of any human action was what the observing scientist said it was, and nothing more. The behaviorist's measuring rod for meaning was as parsimonious as it was mechanical: "The rule, or measuring rod, which the behaviorist puts in front of him always is: Can I describe this bit of behavior I see in terms of 'stimulus and response?' "[42] Watson's definition of these key terms left no doubt that the answer would always be in the affirmative.[43]

Among the manifold responses of the human organism, those which have been learned (conditioned) were to be accounted for as readily as automatic or unconditioned reflexes "on a thoroughly natural science basis without lugging in consciousness or any other so-called mental process."[44] Moreover, the concept of "conditioning" was soon found by Watson to possess an additional and extraordinary, not to say magical, potency—one which made it appear revolutionary indeed in its implications for man and society. Not only did the conditioned reflex furnish a disarmingly simple framework for the explanation of all cognitive activity, but it gave promise of making available the efficient control and manipulation—and ultimately the transformation—of human life. At this point the "objective" spectator of man's behavior suddenly became an active participant in the process under observation. "The interest of the behaviorist is more than the interest of a spectator," said Watson; "he wants to control man's reactions as physical scientists want to control and manipulate other natural phenomena."[45] As man was simply "an assembled organic machine ready to run," so the behaviorist was not just an idle scientist but also an engineer unable to keep from tinkering with the machinery.[46] Observing that chemistry and biology were gaining control of their own subject matter, Watson inquired: "Can psychology ever get control? Can I make someone who is not afraid of snakes, afraid of them and how?"[47] The answer was unequivocal: anything that Nature or Nature's God could do, behaviorism could do better. "The possibility of shaping in any direction is almost endless."[48]

Nor was Watson in any doubt as to the ultimate purposes for which the human organism was to be manipulated, shaped and controlled. They were, in a word, the purposes of society; more specifically, they were the purposes of individual adjustment and efficient performance within the socioeconomic system— which meant, for an American at least, the business system. (Significantly, Watson himself retired from academic life a few years after founding his movement, in order to embark upon a highly successful career as an advertising executive.)[49]

Furthermore, beyond his explicit commitment to the social ethic of the business corporation, Watson shared the conviction of many of his contemporaries of the scientific-management movement (as well as of such later proponents of human engineering as Elton Mayo) that the primordial sin of society in the pre-scientific past had been its tolerance of disorder and its failure to acknowledge the authority of the natural aristocracy of experts.

Indeed, if one were to characterize social experimentation in general during the past 2000 years one would have to call it precipitous, infantile, unplanned, and say that when planned it is always in the interest of some nation, political group, sect or individual, rather than under the guidance of social scientists—assuming their existence. . . . Our own country today is one of the worst offenders in history, ruled as it is by professional politicians, labor propagandists and religious persecutors.[50]

Watson's own prophetic vision was of a world mercifully free not only from politicians, propagandists and religionists, but even from its own history and traditions: a world which, resting statically outside human history, bore a striking resemblance to the City of God. It was to be "a universe unshackled by legendary folk-lore of happenings thousands of years ago; unhampered by disgraceful political history; free of foolish customs and conventions which have no significance in themselves, yet which hem the individual in like taut steel bands."[51] Watson hastened, however, to assure his readers (and followers) that he was "not asking here for revolution; . . . I am not asking for 'free love.' I am trying to dangle a stimulus in front of you, a verbal stimulus which, if acted upon, will gradually change this universe. For the universe *will* change if you bring up your children, not in the freedom of the libertine, but in behavioristic freedom. . . ."[52] The behaviorist, he said in a remarkable footnote, "would like to develop his world of people from birth on, so that their speech and their bodily behavior could equally well be exhibited freely everywhere without running afoul of group standards."[53]

As this makes clear, what the behaviorist desires above all is the opportunity to "develop his world of people from birth on," in order to bring them naturally and automatically into conformity with "group standards." Expressed in such boldly idiosyncratic terms, this portrayal of behaviorism as a precision instrument of social manipulation may seem only a personal aberration of the author rather than an inherent quality of the movement over which he presided. But from its inception the new science of behavior was unmistakably oriented toward practical and remedial ends—not just in the relatively neutral prediction of behavior but in child-rearing, education and the exploitive arts of mass persuasion and "industrial psychology." Behaviorism, in Heidbreder's words, "does not pretend to be a disinterested psychology. It is frankly an applied science, seeking to bring the efficiency of the engineer to bear upon the problem of reform."[54] However, the reform which was contemplated was not the reform of institutions to meet the dynamic needs of men, but the reform of men to meet the static needs of institutions.

Given the behaviorist's fundamental assumption of the irrational and involuntary character of all human behavior— defined as the determinate motion of a stimulus-response machine —it could scarcely be otherwise. Quite apart from the personality and predilections of its founder, behaviorism as a psychological theory would seem to epitomize the outlook of a mass society over which mechanization had taken command—whose ruling norms were those of industrial efficiency and technical proficiency. For such a society the image of the machine furnishes more than a model for scientific investigation; it tends to become the dominant symbol of munificence and beneficence pervading the whole of life: the fountainhead from which all blessings flow. In this way, as Friedrich Juenger has argued, "an advanced stage of technology is accompanied by mechanical theories of the nature of man . . . and all things step by step assume the character of machinery, of a reality understood in terms of machinelike functions."[55]

That the behaviorism of Watson should have been widely heralded, in its heyday of the twenties, as the "authentic" American psychology—despite its sweeping repudiation of consciousness and assault upon reason—furnishes a poignant confirmation of this view. For the student of a later generation, made forcibly aware of the origins and "conditionings" of totalitarianism, there is grim irony in the remark of a commentator of the twenties that behaviorism is especially "adapted to the American temperament because it is fundamentally hopeful and democratic"—not to mention the exclamation of a reviewer of Watson's *Behaviorism*: "Perhaps this is the most important book ever written. One stands for an instant blinded with a great hope."[56]

The great hope which blinded so many upon the advent of behaviorism was, of course, the prospect of extending the hegemony of natural science, already master of the realm of inorganic nature, to the chaotic field of human affairs. For behaviorism surely proclaimed, if not the indefinite perfectibility of man, at least his indefinite malleability and manipulability—which doubtless seemed much the same thing. No less "democratic" in appearance was the behaviorist's stress upon environment as the controlling factor in human development and his accompanying denial of instincts or other fixed hereditary endowments—which seemed to establish the literal truth of the doctrine of natural equality.[57] If, on the other hand, man stood exposed under Watson's analysis as no more than an organic machine, even this fact apparently ennobled rather than enfeebled him—for it only made him more amenable to that triumphant American gospel of engineering and scientific management which to so many carried the hope of mankind for the efficient use of human resources and the rational ordering of social affairs. Within the perspective of scientism, as we have seen before, there is no inconsistency in speaking of the *rational* organization of *irrational* men (or machines); on the contrary, it is exactly the genius of the engineer to make possible the "rationalization" of any mechanical process and so of the individual units which compose it. But within an organized industrial or social process the human units can acquire no more than what Karl Mannheim called "functional rationality"; genuine rationality is attributable only to the system as a whole.[58] Left to their own devices, apart from the cohering system, men are but passive mechanisms subject to the coercion of fleeting stimuli; only in their functional roles within the organization do they take on a measure of meaning and purpose —i.e., of "reason." Strictly speaking, rationality on this view is a property of the *roles* rather than of the beings which fill them; and the roles themselves "make sense" only within the context of the overarching system. Such merely technical rationality, as Juenger has seen, "implies a contempt for human reason as such."[59]

The basis of this contempt for reason and human freedom is that Watsonian behaviorism, like Laplace's System of the World, has no need of that hypothesis. To admit the relevance of reason would be to concede the importance of personal intentions, of conscious purposes and autonomous will—in short, of the capacity not merely to twitch mechanically to the tune of a stimulus but to create and organize activity, to choose among stimuli and control responses. But this would necessitate a sweeping abandonment of all the fundamental tenets of behaviorism. "Since a scientific psychology must regard human behaviour objectively, as determined by necessary laws," the philosopher John MacMurray has argued, "it must represent human behaviour as unintentional."[60]

Whether human conduct is conditioned or unconditioned it remains, on the behaviorist account, wholly determinate and predictable; and, in either event, it is open to manipulation by *re*conditioning ("the possibility of shaping in any direction is almost endless"). It seems extraordinary nowadays that such a doctrine could ever have been construed in any sense as democratic: its blindness to personal intention, its scorn of mind, its denial of any freedom of action or capacity for it, its tacit enlistment in the service of a kind of technocratic efficiency and regimentation—these characteristics of classical behaviorism appear rather to confirm the harsh judgment of Mannheim that it bears an unmistakable resemblance to fascism.[61]

Not only was behaviorism oriented on its practical side to the inhuman use of human beings; it was also, on its theoretical side, less a doctrine of individual psychology than of social psychology. Since, scientifically speaking, all men are "by nature" identical—empty organisms furnished with the same neural and mechanical equipment—they might fairly be treated on the average and in the aggregate, with their superficial differences (the result of minor variations of conditioning) simply ignored. It was this actuarial aspect of behaviorism which was to gain for it perhaps even broader acceptance within sociology than within academic psychology. The militant scientism of Watson and his followers was deeply instrumental in the shaping of that interdisciplinary reductionist impulse recently characterized by Hannah Arendt as "the all-comprehensive pretension of the social sciences which, as 'behavioral sciences,' aim to reduce man as a whole, in all his activities, to the level of a conditioned and behaving animal."[62]

The popularity of behaviorism in its early phase, indeed, was not less extensive among psychologists than among the general public. "For a while in the 1920's, it seemed as if all America had gone behaviorist. Everyone (except the few associated with Titchener) was a behaviorist and no behaviorist agreed with any other."[63] But although the title of behaviorist came to embrace a variety of viewpoints, it seldom meant less than a hard-headed insistence upon objective natural-science methodology, and was commonly associated with "a mechanistic or materialist view of psychology."[64] In fact, there were not a few orthodox behaviorists whose scientistic passion exceeded even that of Watson—among them A. P. Weiss, whose *A Theoretical Basis of Human Behavior* (1925) remains today the most systematic statement of classical behaviorist theory. Weiss was adamant in demanding a place for psychology among the natural sciences, and in urging his colleagues to abandon whatever idiosyncrasies served to alienate them from their fellow scientists.[65] No less ardent in his commitment to rigor and reductionism was Z. Y. Kuo, a well-known disciple of Watson, who became especially exercised about the superstitious residue of "purpose" which he found still lurking in psychological discourse:

> The concept of purpose is a lazy substitute for . . . careful and detailed analysis. . . . The duty of a behaviorist is to describe behavior in exactly the same way as the physicist describes the movement of a machine. . . . This human machine behaves in a certain way because environmental stimulation has forced him to do so.[66]

Aside from the orthodox disciples, however, a number of psychologists who had embraced the behaviorist faith during the twenties subsequently became disenchanted or seriously modified their views.[67] Adding further to the confusion over terminology was the fact that still other theorists, while not unwilling to be called "behaviorists," developed psychological approaches more contradictory to than consistent with the canonical opinions of Watson and Weiss. A notable example was the philosopher and social psychologist George Herbert Mead, who, while describing his own theoretical system as "social behaviorism," disagreed with Watson in virtually all respects other than the adoption of a superficially similar vocabulary.[68] Equally removed from orthodoxy was Edward C. Tolman's system of "purposive behaviorism," whose generous eclecticism comfortably accommodated the insights of Gestalt psychology and even of Freudian psychoanalysis while clinging to a modified framework of stimulus-response behaviorism.[69]

Hull: Behaviorism as Social Science

But if there were deviations and departures from the movement in the thirties, as behaviorism lost the first careless rapture of insurgency and became an established tradition in the textbooks, there was nevertheless a direct line of descent from the doctrine of Watson. Among those who carried on the movement with no appreciable slackening of its original commitment and zeal were two distinguished experimentalists, Clark L. Hull of Yale and B. F. Skinner of Harvard, whose contributions not only to behaviorist psychology but to the broader development of behavioral science deserve particular attention.

The extent of Hull's influence is indicated by the fact that, for a full generation prior to his death in 1932, he was the theoretical center of a distinguished group of social scientists at the Yale Institute of Human Relations who represented such varied disciplines as sociology, psychiatry and anthropology, and whose combined production has given the most concerted and sustained impetus to the shaping of the behavioral-science approach.[70] Even the name itself may well have been coined by Hull, who referred to his most influential work, *Principles of Behavior* (1943), as "a general introduction to the theory of the behavioral (social) sciences," conceived and carried out "on the assumption that all behavior, individual and social, moral and immoral, normal and psychopathic, is generated from the same primary laws."[71]

No less than Watson, Hull wore his mechanistic philosophy (composed of equal parts of objectivism and reductionism) on his sleeve. On no point was he more insistent than that a genuine theory of human behavior must require the expulsion of all traces of what he called "anthropomorphic subjectivism"—i.e., any betrayal of the presence of a valuing human observer. "A genuinely scientific theory no more needs the anthropomorphic intuition of the theorist to eke out the deduction of its implications than an automatic calculating machine needs the intuitions of the operator in the determination of a quotient, once the keys representing the dividend and the divisor have been depressed."[72] Human behavior was to be regarded as a purely automatic cyclical operation, each cycle beginning with the rise of a demonstrable "need" and ending with its abolition or reduction.[73] "So-called purposive behavior" (i.e., action involving the appearance of intelligence and intent), while granted a kind of epiphenomenal reality, was ultimately to be accounted for by deriving it as secondary phenomena from "more elementary objective primary principles."[74] The very notion that human behavior could not be reduced in all cases (individual or social, moral or immoral) to automatic mechanical processes, identical for men and animals, was curtly dismissed by Hull as a "defeatist attitude" and a "doctrine of despair."[75]

Under the heading, "A Suggested Prophylaxis Against Anthropomorphic Subjectivism," Hull warned that even the most disciplined and wary of behavioral scientists, seeking to gain the proper detachment from their subject-matter, were in danger of falling prey to the "seductions" of anthropomorphic subjectivism. Two alternative prophylactics were recommended against this peril: a psychogenetic model or, better yet, a purely mechanical model.

One aid to the attainment of behavioral objectivity is to think in terms of the behavior of subhuman organisms, such as chimpanzees, monkeys, dogs, cats, and albino rats. Unfortunately this form of prophylaxis against subjectivism all too often breaks down when the theorist begins thinking what he would do if he were a rat, a cat, or a chimpanzee; . . .

A device much employed by the author has proved itself to be a far more effective prophylaxis. This is to regard, from time to time, the behaving organism as *a completely self-maintaining robot,* constructed of materials as unlike ourselves as may be.[76]

There were two marked advantages to be found in what Hull called the "robot approach": first, it was a bracing mental exercise for the scientist to tackle the technical problems in behavior dynamics involved in the design of a truly "self-maintaining robot"; and, second, the robot concept effectively removed any tendency to "reify" behavior functions by attributing other than mechanical causes to the conduct under observation. Through the reduction of human behavior to the automatic level of the robot and the consequent attainment by the observer of disinterested neutrality, it became possible for Hull to argue that the difference between the physical and behavioral sciences is "one not of kind but of degree," and that the relative backwardness of the behavioral sciences is due mainly to "the difficulty of maintaining a consistent and rigorous objectivism."[77] The notorious recalcitrance of human subjects to such reductive analysis was, said Hull, a "source of regret" but no cause for chagrin; it would take time to surmount the "historical contamination of behavior science by metaphysical speculation,"[78] but eventually all the underlying laws and automatisms would be known and the troublesome ghost forever exorcised from the machine. In the meantime, the progress of behavioral science must depend upon faithful adherence to the conception of the human organism as, like other forms of life, a self-maintaining mechanism—"defined as a physical aggregate whose behavior occurs under ascertainable conditions according to definitely statable rules or laws."[79]

Throughout his later career the principal goal of Hull's research was the articulation of a systematic "mathematico-deductive" theory of behavior—which was to include *all* behavior, animal and human, individual and social, political and moral. By

1943 he had isolated some sixteen basic principles or laws of behavior, from which it was hoped to "deduce an extensive logical hierarchy of secondary principles which will exactly parallel all of the objectively observable phenomena of the behavior of higher organisms; such a hierarchy would constitute a systematic theory of all the social sciences."[80] On this basis it was anticipated that further treatises, of steadily mounting abstraction and quantitative sophistication, would emerge embracing all aspects of behavioral science: first a general theory of individual behavior, to be followed by a general theory of "social behavior." With the progressive elaboration of all the "various subdivisions and combinations of these volumes" there would emerge a comprehensive body of theoretical materials dealing with all aspects of mammalian behavior, including that of the human mammal. Something of the encyclopedic sweep of this projected behavioral science is conveyed by the following complex sentence:

> Such a development would include volumes devoted to the theory of skills and their acquisition, of communicational symbolism or language (semantics); of the use of symbolism in individual problem solution involving thought and reasoning; of social or ritualistic symbolism; of economic values and valuation; of moral values and valuation; of aesthetic values and valuation; of familial behavior; of individual adaptive efficiency (intelligence); of the formal educative processes; of psychogenic disorders; of social control and delinquency; of character and personality; of culture and acculturation; of magic and religious practices; of custom, law, and jurisprudence; of politics and government; and of many other specialized behavior fields.[81]

Nor was this all. "As a culmination of the whole, there would finally appear a work consisting chiefly of mathematics and mathematical logic."[82] The crowning apex of all these pyramidal contributions—the ultimate answer to the riddle of life and conduct—would consist of a master set of equations worked out by "the incomparable technique of symbolic logic." Through an inexorable winnowing process, the whole paraphernalia of signs and symbols would be steadily reduced until there had been "derived theorems paralleling all the empirical ramifications of the so-called social sciences. Also there should be derivable large numbers of theorems concerning *the outcome of situations never yet investigated....*"[83]

To the human intelligence which would finally comprehend all these variables within the single master formula, nothing, evidently, would be uncertain. The cosmic vision of Hull and his associates was scarcely inferior to that of Laplace himself, and indeed was founded on the same faith: an unswerving acceptance of the indefinite perfectibility and universal applicability of exact scientific methods, as well as of the eventual attainment of something like certainty through discovery of the mechanical laws underlying reality and their articulation in the incomparable language of mathematics.[84]

It was a luminous vision, if somewhat clouded by the anticipation of bitter-end resistance on the part of jealously opposed interests.[85] Nevertheless it seemed to Hull that the next hundred years would witness an unprecedented fruitfulness and progress—due mainly to the heartening disposition, "at least among Americans, to regard the 'social' or behavioral sciences as genuine natural sciences rather than as *Geisteswissenschaft*."[86] If this hopeful tendency should prevail, there was every reason to expect that the behavioral sciences would presently display "a development comparable to that manifested by the physical sciences in the age of Copernicus, Kepler, Galileo, and Newton."[87] In that divine event "the so-called social sciences will no longer be a division of *belles-lettres*; anthropomorphic intuition and a brilliant style" would no longer suffice to beguile the unwary and hold back the dawn. Indeed, the only expository "style" with much likelihood of survival would be that of abstract mathematical symbolism, together with the rediscovered magic of pure number.

Progress in this new era will consist in the laborious writing, one by one, of hundreds of equations; in the experimental determination, one by one, of hundreds of the empirical constants contained in the equations; in the devising of practically usable units in which to measure the quantities expressed by the equations; in the objective definition of hundreds of symbols appearing in the equations; in the rigorous deduction, one by one, of thousands of theorems and corollaries from the primary definitions and equations; in the meticulous performance of thousands of critical quantitative experiments and field investigations designed with imagination, sagacity, and daring . . .[88]

But this "great task," clearly, could be no more than begun by the present generation. The real hope must lie with the oncoming youth, to whom would belong the "thrill of intellectual adventure" and ultimate fulfillment of the vision. "Perhaps," wrote Hull at the conclusion of his major book, *"they will have the satisfaction of creating a new and better world, one in which, among other things, there will be a really effective and universal moral education."*[89]

Here once again—as in the case of the earlier prophets of positivism and behaviorism—there arises the inherent paradox of scientism. On the basis of a rigorously value-purged objectivity— for which morality is irrelevant to the point of incompetence—the natural science of human behavior looks forward to a "new and better world" suffused with the light of moral wisdom. Nor can it be said that this access of ethical consciousness on the part of Hull was only a rhetorical flourish.[90] His final volume, *A Behavior System* (1952), was partly given over to the elaboration of a new "natural science of ethics" aimed at the quantitative description of "all kinds of behavior of organisms, whether generally characterized as good, bad, or indifferent," and including specifically the making of moral judgments.[91] If this reductionist formulation was not sufficient to meet the overwhelming question—i.e., can science tell us what is good?—Hull did not hesitate to confront that issue directly. All statements of ethical theory, of what men ought or ought not to do, he wrote, "occupy a scientific no-man's land; which is practically equivalent to saying that such statements are scientific nonsense."[92] But while the traditional science of ethics, concerned with *evaluation*, can never be more than a "pseudo-science," there is nothing to prevent the development of a "true natural science of moral behavior" concerned not with evaluation but with the derivation and quantitative analysis of moral actions —and, above all, with "the application of science in the determination of the most effective means of attaining values of all kinds, ethical or otherwise."[93]

With this statement the Hullian vision of a new and better world fades into a familiar obscurity. The only originality left in it is that of an ethical nihilism which simultaneously repudiates as nonsense all moral beliefs derived on extrascientific grounds while it rejects as impossible all efforts by science to validate or correct them. And the only innovation left in it is that, "as quantitative behavioral symbolic constructs are gradually perfected," all ethical systems grounded on nonscientific (metaphysical) foundations will be perforce abandoned: "Then the theory of value will cease to be a division of speculative philosophy and will become a *bona fide* portion of natural science."[94] And then also, one might add, the underlying values upon which the entire enterprise of science is predicated—having lost even the discredited authority of speculative philosophy—will become unexamined and unchallengeable "givens," to which behavioral scientists will be simultaneously indifferent and committed. They will be indifferent because all moral values are "scientific nonsense"; but they will also be bound because without some such values no social science, however behavioristic, can perform the task which Hull himself has recognized as definitive: namely, "the determination of the most effective means of attaining values of all kinds"—nonsensical or otherwise.

It is at least plausible to argue that the ethical cul-de-sac into which Hull's natural science of behavior has led him is the inevitable outcome of the mechanical model on which his image of human nature is constructed. A robot may perhaps be "self-maintaining," but it remains nevertheless a robot—a *creature* in the literal sense of the term. At the least it requires a designer, if it does not also demand an engineer to check its oil and rationalize its performance. In short, the robot can never be a *subject*, an end in itself; it acquires its simulated purpose, its artificial reason for "being," wholly from the outside.[95] Regarded in this mechanical perspective, man also is reduced to the status of an object, a passive

means to some imposed external end. Scientific interest in his
behavior thus naturally comes to focus upon its prediction and
control (no behaviorist would say its "understanding"), precisely
as in the detached analysis of physical phenomena. Within such a
framework it is entirely appropriate to say that the human organism
cannot be valued "for itself," because in fact it has no *self:* no
autonomy, no unique identity, no genuinely independent existence.[96] Its fate is either to drift through life in a meaningless cycle
of titillation-and-response, or to be conscripted in the service of a
superior power.

There need be nothing consciously sinister, of course, in
such manipulation. Indeed, given the behaviorist thesis of the
"empty organism," it is rather an act of charity to seek its rationalization and repair through human engineering. And those behavioral scientists who have most earnestly undertaken theoretical
programs for the social and moral rehabilitation of man have—
from the time of Comte onward—had only his best interests at
heart.

Skinner: Behaviorism as Cultural Design

This is, perhaps, most conspicuously the case with B. F.
Skinner. Where Hull had devoted the bulk of his career to elaborating the framework of a "mathematico-deductive" and hence
predictive science of behavior, Skinner—who has succeeded Hull
as perhaps the most influential of contemporary behaviorists[97]—
has gone further to contemplate an applied science aimed frankly at
the manipulation and control of human conduct. More plainly even
than Watson's, his writings reveal the deep urge of the behaviorist
not only to observe men behaving but to *make* them behave.

The potential authority of a rigorous science of behavior
was well appreciated by Skinner in his earliest major work, *The
Behavior of Organisms* (1938), in which he declared that "It is
largely because of its tremendous consequences that a rigorous
treatment of behavior is still regarded in many quarters as impos-

sible."[98] While paying tribute to Pavlov and Watson as pioneers in the development of such a behavioral science, Skinner insisted that they had not gone far enough; both had risen above the sterile tradition of "mentalistic" psychology, but they had also retained some of its superstition by relying upon the equally internal (subjective) processes of the nervous system. What was needed was a more radical and literal behaviorism: one that would limit itself strictly to "what an organism is doing—or more accurately what it is observed by another organism to be doing."[99] By "behavior" Skinner proposed to mean "simply the movement of an organism or of its parts in a frame of reference provided by the organism itself or by various external objects or fields of force."[100] A theoretical system restricted so rigidly to observables can have no place for concepts dealing with psychic events or inner states, and its author promptly banished as irrelevant such archaic terms as "meaning," "intent," and "understanding."[101]

Given the actual subjects of his laboratory investigations, there would seem to have been little need for such caution. Where Hull, after some experimentation, had settled upon the robot as the ideal model for the study of organic behavior, Skinner has consistently expressed a preference for the white rat (and, to some degree, the pigeon). Although less accommodating than the fully automatic robot, the rodent, he has said, at least "has the advantage over man of submitting to the experimental control of its drives and routines of living."[102] It is instructive to note the degree of difference which Skinner admits to exist between man and the rat: "It differs from man in its sensory equipment (especially in its poorer vision), in its reactive capacities (as of hands, larynx, and so on), and in limitations in certain other capacities such as that for forming discriminations."[103] That is the full extent of the difference. Still more succinctly, at the conclusion of the same study, Skinner hazards the guess that "the only differences I expect to see revealed between the behavior of rat and man (aside from enormous differences of complexity) lie in the field of verbal behavior."[104]

As one critic has remarked concerning this last passage, the differences implied by the term "verbal behavior" may well be interpreted to cover a very wide ground, since most of human behavior depends upon the use and understanding of language.[105] But it appears likely, in light of the above quotations, that Skinner intended to minimize rather than to magnify the differences between the two species. This impression is reinforced by the obvious willingness of Skinner and others to engage freely in theoretical discussion of human behavior on the basis of laboratory work carried out predominantly or exclusively on nonhuman subjects. Thus Lafitte points out that Skinner's choice of the white rat because it is "similar to man" suggests the confusion found in many elementary texts, in which "the results of animal experiments (e.g., on learning and motivation) are presented in indiscriminate combination with findings derived from persons, regardless of whether or not the animal experiments have been repeated with persons."[106]

There is a curious lack of consistency in this attitude on the part of behaviorists toward the objects of their research. On the one hand they are seldom reluctant to attribute to human behavior the mechanisms and automatisms found in the study of nonhuman subjects, whether rodents or robots; on the other hand they are almost obsessively concerned over the error of imputing "anthropomorphic" tendencies to any organisms whatsoever—including human organisms. For Skinner as for Hull, "anthropomorphism" is the supreme peril to be avoided at all costs; but neither authority has evidenced any concern over the alternative peril of "mechanomorphism."[107] The paradox which this behaviorist bias suggests has been crisply pointed out by Joseph Wood Krutch:

What many psychologists and social scientists seem to be doing is to denounce as a kind of "anthropomorphism" every attempt to interpret even human behavior on the assumption that men are men; to insist that we should proceed as though they were mere animals at most, even if not mere machines in the end. But how can Anthropos be understood except in anthropomorphic terms? Or should we assume that the mechanomorphic error is not really an error at all?[108]

For the consistent behaviorist, to be sure, these questions pose no serious difficulty; the very meaning of "anthropomorphic" is simply that of "subjective," and there is patently no place for the subjective in an objective science of behavior. It is important, once again, to recognize that the interest of such a behavioral scientist is not in the *understanding* of human conduct (whatever, as he would say, that may mean) but simply in its prediction and control. His fundamental commitment—if not the only obligation he would care to acknowledge—is to his version of scientific method and his vision of scientific truth.

If this inference seems unwarranted, it is only necessary to call in evidence the considered opinions and prescriptions of Skinner, as they have been expressed in an unusual corpus of works ranging from a comprehensive treatise on *Science and Human Behavior* to a remarkable utopian novel, *Walden Two*. In these writings—most systematically in the former, most imaginatively in the latter—Skinner has sought with an admirable candor and absence of equivocation to outline a workable program for the effective control of human behavior. It is a program which, if seriously undertaken, may be expected literally to transform the character of man as well as of his cultural environment.

A Program of Behavioral Control

The first step in the creation of an applied science of human behavior, we are told, is to recognize that it is both feasible and urgent; indeed, it may be our only hope. But there is an important condition attached to this salvation by science: "If we are to enjoy the advantages of science in the field of human affairs, we must be prepared to adopt the working model of behavior to which [such] a science will inevitably lead."[109] What that working model indicates, of course, is the thoroughly determinate character of all human action; in short, it is the model of man as an organic machine. The commitment to rigorous law and order everywhere in the universe is, for Skinner, "a working assumption which must be adopted at the very start. *We cannot apply the methods of science to a subject matter which is assumed to move about capriciously*."[110] There is no room for puckish or disorderly conduct on the part of any of Nature's creatures. For "science not only describes, it predicts. It deals not only with the past but with the future." More than that: the scientist cannot be confined to the passive role of the prophet; he must be the prophet armed. To control man is to control the future.

> To the extent that relevant conditions can be altered, or otherwise controlled, the future can be controlled. If we are to use the methods of science in the field of human affairs, we must assume that behavior is lawful and determined . . . that what a man does is the result of specifiable conditions and that once these conditions have been discovered, we can anticipate and to some extent determine his actions.[111]

This manipulatory prospect, Skinner readily admits, "is offensive to many people"; it runs counter to that old tradition which regards man as a free agent somehow capable of carving out his own career and of resisting the iron laws of natural causation. But the issue must be squarely faced, for it has serious practical consequences. "*A scientific conception of human behavior dictates one practice, a philosophy of personal freedom another. . . .* The present unhappy condition of the world may in large measure be traced to our vacillation."[112] It is therefore essential to be clear about the alternatives: on the one hand there is the traditional view of human nature, "the conception of a free, responsible individual [which] is embedded in our language and pervades our practices, codes and beliefs"; on the other hand is the "new and strange" viewpoint of science—the conception of man as unfree and irresponsible—whose power and promise have yet to gain equivalent recognition.[113]

If we are to harness this scientific potential, Skinner insists, we must be prepared first of all to give up the habit of viewing behavior "in terms of an inner agent which lacks physical dimensions and is called 'mental' or 'psychic.' "[114] This naive habit of looking *inside* the organism for answers has kept us from seeing the actual causes, the real independent variables, which have been there under our noses all along—i.e., those that lie outside the organism in its environment and environmental history. Being palpable and measurable, like all scientific objects, "they make it possible to explain behavior as other subjects are explained in science."[115]

All of this must of course be accomplished, Skinner points out, "strictly within the bounds of a natural science. We cannot assume that behavior has any peculiar properties which require unique methods or special kinds of knowledge."[116] Everything in the behaving system must be reducible to physical terms, without any lapses into subjective anthropomorphism in order "to avoid the labor of analyzing a physical situation by guessing what it 'means' to an organism or by distinguishing between the physical world and a psychological world of 'experience.' "[117] For such a science the ideal materials would be those derived from experimental studies of human behavior; but unfortunately direct work on human beings "is sometimes not so comprehensive as one might wish"—for obvious unscientific reasons.[118] In these frustrating circumstances our information must be largely drawn from experimental studies of animals below the human level—studies which, if not quite conclusive, are perfectly adequate and appropriate.[119]

Above all, according to Skinner, if we are to construct a genuine science of behavior we must be thoroughgoing about it; we cannot have our science and evade it too. "If we are to further our understanding and to improve our practices of control, we must be prepared for the kind of rigorous thinking which science requires."[120] Just what kind of thinking that implies is made clear by Skinner in the course of a discussion of "Personal Control" (i.e., *control of other persons*), in which he describes with startling detachment the comparative advantages and drawbacks of such time-honored techniques of behavioral control as, for example, killing, "the use of handcuffs, strait jackets, jails, concentration camps, and so on."[121] Skinner's conclusion is equally forthright:

The countercontrol exercised by the group and by certain agencies may explain our hesitancy in discussing the subject of personal control frankly and in dealing with the facts in an objective way. But it does not excuse such an attitude or practice. *This is only a special case of the general principle that the issue of personal freedom must not be allowed to interfere with a scientific analysis of human behavior.* . . . We cannot expect to profit from applying the methods of science to human behavior if for some extraneous reason we refuse to admit that our subject matter can be controlled.[122]

If "personal freedom" must not be allowed to interfere with the scientific management of behavior—and so is, to say the least, disregarded as a relevant value—what then *are* the purposes for which science is to be harnessed and its controls exercised? Skinner's answer to the ethical question is not quite that of Comte (the scientific society), nor that of Watson (the *status quo*), nor yet that of Hull ("nonsense"). The ultimate criterion of the new science of behavior is simply that of *survival*—not the survival of the individual but of the "culture." Whatever has survival-value for the social culture is "good"; whatever does not contribute to its survival is "bad."[123]

A rigorous science of behavior, then, "leads us to recognize survival as a criterion in evaluating a controlling practice."[124] It does so by taking into account its long-range, and hence most important, consequences for the social group—as opposed to such obvious short-range considerations as "happiness, justice, knowledge, and so on." These easy virtues may now at last be seen in a realistic perspective: "A scientific analysis may lead us to resist the more immediate blandishments [sic] of freedom, justice, knowledge, or happiness in considering the long-run consequences of survival."[125]

It should perhaps be said here that Skinner is fully cognizant of the "problem of control" posed by the rapid growth of behavioral science; he emphasizes that "the effect upon human affairs will be tremendous."[126] What sort of effect is this to be? How are the controls themselves to be controlled? First of all, says Skinner, we must not deny the plain reality of manipulation; it goes on all the time, and everybody does it.

One proposed solution is to insist that man is a free agent and forever beyond the reach of controlling techniques. It is apparently no longer possible to seek refuge in that belief. . . . A doctrine of personal freedom appeals to anyone to whom the release from coercive control is important. But behavior is determined in noncoercive ways; and as other kinds of control are better understood, *the doctrine of personal freedom becomes less and less effective as a motivating device and less and less tenable in a theoretical understanding of human behavior. We all control, and we are all controlled.* As human behavior is further analyzed, control will become more effective. Sooner or later the problem must be faced.[127]

Nor is it any answer to reject the new "opportunity to control" which science has made available; for if one of us should let this cup pass from him, another is sure to seize it.[128] It is all very well to say, as certain psychotherapists do say, that the human person has within himself the means to find his own solutions and therefore that the therapist need not take action for him. But if the individual is the product of the wrong kind of culture, or has otherwise been poorly conditioned, "no acceptable solution may be available 'within himself.' "[129] Someone else, who knows what is the *right* kind of culture, will have to decide for him. On the level of politics, all that a free society really means anyway is that "the individual is controlled by agencies other than government"—not that he is ever "free" in the sense of possessing effective *self*-control and the capacity for self-determination.[130]

Skinner's own ultimate solution lies in a curiously paternalistic theory of government which appears to bestow upon the rulers all the rationality it has denied to the ruled. While conceding that there may be some risks in a monopoly by government of the scientific weapons of manipulation, Skinner nevertheless contends that "the ultimate strength of a controller depends upon the strength of those whom he controls."[131] What this odd proposition seems to mean is that it is always in the interest of government— any government—to minimize resentments which might lead to "counter-controls" against itself, and so to maximize the well-being of its subjects. (In fact, for Skinner there is really not much difference in this respect between one form of government and another. "Under present conditions of competition, it is unlikely that a government can survive which does not govern in the best interests of everyone.")[132]

Once this extraordinary precept is acknowledged, it follows that behavioral science can play a key role in determining what those "best interests" are. This possibility arises from the guiding standard of "survival"; by demonstrating that programs of welfare and happiness "pay off" in the practical terms of international competition and rivalry, we might make a really convincing scientific case for such policies. For example, with the aid of science we can "evaluate the use of physical force by considering the ultimate effect upon the group" *employing* it—without bothering about its effects upon the victim. "Why should a particular government not slaughter the entire population of a captured city or country?" Skinner asks. "It is part of our cultural heritage to call such behavior wrong and to react, perhaps in a violently emotional way, to the suggestion."[133] But as behavioral scientists we need no longer have recourse to such outbursts; we can demonstrate dispassionately through charts and figures that *such a practice would weaken the government.*" To take another example, rather than objecting to slavery in the old way, "because it is 'wrong,' or because it is 'incompatible with our conception of the dignity of man,' an alternative consideration in the design of culture might be that slavery reduces the effectiveness of those who are enslaved and has serious effects upon other members of the group."[134] Thus the science of human behavior produces its own appropriate moral standard—that of "group survival"—which, being wholly objective and empirically verifiable, is clearly to be preferred to all those emotional and utterly unquantifiable "moral laws" which have preceded it down the dark corridors of prescientific history.[135]

But if this scientific morality is sometimes to be found in support of traditional notions of freedom or justice, it is not necessarily so. On the contrary, Skinner tells us, that central tradition of Western thought which has cherished the essential dignity and liberty of the individual is, quite simply, no longer tenable in the face of modern scientific knowledge of the nature of man.

> The use of such concepts as individual freedom, initiative, and responsibility has, therefore, been well reinforced. When we turn to what science has to offer, however, we do not find very comforting support for the traditional Western point of view. *The hypothesis that man is not free is essential to the application of scientific method to the study of human behavior.* The free inner man who is held responsible for the behavior of the external biological organism is only a prescientific substitute for the kinds of causes which are discovered in the course of a scientific analysis. All these alternative causes lie *outside* the individual. . . . These are the things which make the individual behave as he does. For them he is not responsible, and for them it is useless to praise or blame him.[136]

To be sure, such a scientific view "is distasteful to most of those who have been strongly affected by democratic philosophies." But there is no help for it; the old way of thinking is doomed. Behavioral science, it now appears, is by no means a neutral instrument to be used however we see fit; once its mechanism is wound up and set going, it points its own direction and follows its own daemon. The creature thus comes to dominate its creator, *and this is as it should be;* our deepest beliefs and oldest traditions are helpless to check or divert its forward march. It would seem that all that is left for modern man is to regard the movement of science—specifically, of behavioral science—not as a juggernaut but as a bandwagon, and to climb aboard while there is time.

> . . . it has always been the unfortunate task of science to dispossess cherished beliefs regarding the place of man in the universe. . . . If science does not confirm the assumptions of freedom, initiative and responsibility in the behavior of the individual these assumptions will not ultimately be effective either as motivating devices or as goals in the design of culture. . . . The highest human dignity may be to accept the facts of human behavior regardless of their momentary implications.[137]

Walden Revisited: A Behavioral Utopia

The scientific blueprint for a happily controlled and conditioned society which has been spelled out systematically in *Science and Human Behavior* is also available, for the nonacademic reader, in a popular version. Skinner has written a science-fiction novel, *Walden Two*, which despite minor concessions to the demands of plot and romance is basically faithful to the cultural grand design and wholly serious in its intention. Since this futuristic fable has been acutely analyzed by Joseph Wood Krutch and others,[138] it may be sufficient here to review a few of the suggestive ways by which its author has reconstructed his behavioral theories and political ideals.

Walden Two is a utopian community, somewhere in the eastern United States, founded by an experimental psychologist named Frazier whose pronouncements throughout the book are a close paraphrase of the views expressed elsewhere by his creator.[139] The model society is strictly a product of "behavioral engineering," carried to the point of the virtual extinction of all inconvenient emotions and anti-social impulses. There are no moral problems to be found here, nor any serious issues of choice: the residents, or inmates, are instead conditioned from birth to make the "right choices" automatically without the burden of conscious decision or volition. The government of the place, while ostensibly in the hands of a Board of Planners, is actually controlled by a group of Managers—"specialists in charge of the divisions and services of Walden Two."[140] (Among them, for example, are a Manager of Personal Behavior and a Manager of Cultural Behavior.) The community of subjects plays no part in the selection of these rulers, who, as it turns out, co-opt themselves somewhat on the order of a corporate board of directors. There is little or no political participation by the general population even on lesser levels; active control is gladly left to the experts. The constitution of the model society is embodied in a Code whose provisions may be altered only by the Planners and are not subject to public discussion (a formal caution which might seem superfluous for a community already thoroughly conditioned to passivity). The moral of Skinner's tale, as spelled out at the end by its narrator, is plainly that our own society should adopt forthwith the social, political and behavioral principles brought to such perfection by the Managers of Walden Two. "What was needed," the book concludes,

> was a new conception of man, compatible with our scientific knowledge, . . . But to achieve this, education would have to abandon the technical limitations which it has imposed upon itself and step forth into a broader sphere of human engineering. Nothing short of the complete revision of a culture would suffice.[141]

Here, then, in textbook and storybook, is the outline of a total science of human behavior—at once descriptive and prescriptive, pure and applied—which has emerged from the experiments and scientific premises of a leading contemporary exponent of behaviorism. Skinner's behavioral science is, if anything, more uncompromising in its abolition of mind and purpose—not to mention the aspirations of what a prescientific age was wont to call the human spirit—than the classical doctrine of Watson himself. It is an altogether explicit disavowal of the freedom and responsibility, as well as the moral and political primacy, of the human person; and it is, correspondingly, an avowal of the superior value and importance of "society." Skinner's projection of the scientific utopia—a culture designed and engineered entirely on the rational counsel of behavioral science—is an unflinching extrapolation from the data, the methods and the underlying world view of the scientific mechanist. Given these assumptions about human nature and, more specifically, about the involuntary springs of action; given this expert knowledge concerning the new techniques, whether coercive or subtly palatable, of human conditioning and control; given the urge to practice these skills upon humanity in the service of a vision of perfection and efficiency—given all this, it is hard to see how Skinner's prescriptive "design for a culture" might be improved upon. It is not too high a compliment indeed to say that his program of applied behavioral science possesses much the same moral urgency, the same inexorable logic and persuasive power—and the same totally authoritarian personality—as Hobbes's *Leviathan*.

From the Hobbesian commonwealth to the Elysian fields of Walden Two, the behavioral science of psychology has given rise to a recurrent political dream of redemption and reconstruction. But it is not only psychologists who have been drawn by this compelling vision. We turn next to the science of politics itself for a related perspective on the nature and destiny of behavioral man.

CHAPTER III

The Manipulated Society:
Politics as the Science of Behavior

The scientist, as such, has no ethical, religious, political, literary, philosophical, moral, or marital preferences. . . . As a scientist he is interested not in what is right or wrong or good or evil, but only in what is true or false.

—Robert Bierstedt

There is nothing, then, in the proposals of this book or in the social order which it envisions which cannot be supported by the adherents of every religion, every political and every economic faction interested in the efficient and economical achievement of whatever objectives they pursue.

—George A. Lundberg

T HE PSYCHOLOGICAL SYSTEM of behaviorism, as the preceding chapter has sought to show, is in major respects hardly a system of individual psychology at all. By virtue of its mechanistic assumptions, its actuarial bias and its narrowly utilitarian objectives, behaviorism has tended to focus attention upon the average and aggregate characteristics of human beings and to regard the person as little more than a social atom—for all "behavioral" purposes equivalent if not identical to other units of the same generic class.

Given this objective and statistical emphasis, it is only natural that the appeal of behaviorism should soon have extended beyond psychology to the social sciences. By the mid-1930's its progress had led the European sociologist Karl Mannheim to warn of the consequences for his own field. Observing that behaviorism had strongly reinforced the tendency to concentrate on "entirely externally perceivable reactions" and to "construct a world of facts in which there will exist only measurable data," he uttered a significant prophecy:

> It is possible, and even probable, that sociology must pass through this stage in which its contents will undergo a mechanistic dehumanization and formalization, just as psychology did, so that out of devotion to an ideal of narrow exactitude nothing will remain except statistical data, tests, surveys, etc., and in the end every significant formulation of a problem will be excluded.[1]

In point of fact, the effects of this "mechanistic dehumanization" had begun to be felt within American sociology at least a decade before Mannheim warned of them. Perhaps the clearest reflection of the behaviorist impact is to be seen in the later writings of Franklin H. Giddings, who came to regard sociology as merely the "psychology of society"—more exactly, as the scientific study of "pluralistic behavior," defined as the predictable response of a "plurel" or group to objective stimuli.[2] With Giddings' influential work there was begun what one historian has termed the neo-positivist movement in sociology, the principal ingredients of which were "quantitativism, behaviorism, and positivist epistemology."[3] Whatever the precise accuracy of this description, it is plain that by the thirties a school of thought closely parallel to psychological behaviorism had become established in sociology, which could enter a claim as well authenticated as any to the status and title of behavioral science.

What is most notable about this development, from the standpoint of our inquiry, is that today the heaviest concentration of behavioralists in the field is to be found within the special province which has become known as Political Sociology, or the Sociology of Politics; and more specifically in the study of public opinion, voting behavior, and the persuasive aspects of the mass media. (It is also true, of course, that these standard interests of political sociology have attracted students from a wide variety of traditions and perspectives, some of them outspokenly opposed to the attitudes of the behavioralists.)[4] It is more than coincidence that the phenomena of voting, public opinion and mass persuasion are also the dominant concerns of one of the newest subdivisions within political science: i.e., the study of political behavior.[5] Although both of the parent disciplines, sociology and political science, have been dealing with these matters in their available contexts for upwards of a century, it is the distinctive contribution of behavioral scientists to have forsaken the traditional bases of inquiry in favor of staking out a new and exclusive province of their own.[6] The general boundaries of that province have seldom been more graphically described than in the following statement by Bernard Berelson, summarizing a quarter-century of progress in his own field of public opinion.

> Put together, these differences spell a revolutionary change in the field of public opinion studies: the field has become technical and quantitative, a-theoretical, segmentalized, and particularized, specialized and institutionalized, "modernized" and "group-ized" —in short, *as a characteristic behavioral science*, Americanized. Twenty-five years ago and earlier, prominent writers, as part of their general concern with the nature and functioning of society, learnedly studied public opinion not "for itself" but in broad historical, theoretical, and philosophical terms and wrote treatises. Today, teams of technicians do research projects on specific subjects and report findings. Twenty years ago the study of public opinion was part of scholarship. Today it is part of science.[7]

The New Retreat from Politics

All of the important implications of social and political behaviorism, as well as its underlying assumptions, are to be found in Berelson's forthright definition of the behavioral science of public opinion. Basic to them all is the familiar commitment of scientism to the ideal of rigorous objectivity, which imposes a corresponding neutrality toward any and all values that may arise in the process of inquiry—and which, accordingly, requires the thorough purgation by the observer of his own value-preferences, or "biases," prior to entering the laboratory. "As a scientist he is interested not in what is right or wrong or good or evil, but only in what is true or false."[8]

Not only do such scientists of behavior, in pursuit of the grail of objectivity, seek to preserve a chaste detachment from all that is value-laden in their field of investigation. Their "methodological inhibition," as the late C. Wright Mills termed it (Mannheim called it "methodological asceticism"), often operates to narrow the field itself and to justify an avoidance of troublesome areas where issues of consequence are being decided—that is, where values are in conflict and human passions likely to be involved.[9] It might be supposed that behavioral scientists, facing the challenge of such human problems and recognizing the inadequacy of their analytic tools, would cheerfully admit their failure and turn to the discovery of new and appropriate techniques. No doubt there are instances where this soul-searching has occurred; more commonly, however, what would seem to be a weakness has been converted by the logic of methodolatry into a virtue. The same criteria which guide the "teams of technicians" to the fashionable margins of research also militate against the selection of those classic subjects which once commanded the attention of serious social scientists as a matter of course.[10] Thus, for a striking example, in an authoritative text on research methods in the social sciences (under the appropriately severe heading, "Delimitation of the Scope of Inquiry"), the author cites with approval the warnings of the influential sociologist W. F. Ogburn against attempts to carry research into areas not readily amenable to the demands of scientific rigor.

> Ogburn warns against choosing topics for social research on which—with the present development of the social sciences—there are no valid data or where the basic underlying researches are exceedingly difficult and have not been made. We should also consider the degree of accuracy or approximation essential for the demands of science . . . potential data likely to be strongly colored by emotions may lead to distortions and inaccuracies.

International relations, strikes and lockouts, poverty and riches are examples of topics heavily weighted with emotions and should, therefore, be carefully considered both from the standpoint of feasibility of obtaining accurate and reliable facts and methods of approach.[11]

What this advice means, in blunt terms, is that social scientists whose concern is properly with rigor and accuracy—with measurable certainty and unambiguous prediction—should hang their clothes on a safely dead limb and not go near the water. That the injunction to shrink from issues involving emotion has been widely followed over recent years is a commonplace of criticism by nonbehavioral scientists. It is illustrated in the proliferation of scope-and-method studies over the past decade or so; in the conspicuous preference of "abstracted empiricists" for the external mechanics and minutiae of political behavior,[12] and in the rise to prominence of what Barrington Moore has termed a "new scholasticism" of formal theory centering on the construction of generalized political or social systems kept carefully insulated from empirical reality.[13]

In effect, there would appear to be three separate approaches to the study of politics available to the value-free behavioral scientist—each of which, it can be argued, is in fact a retreat from politics. First, he may choose to concentrate upon those mechanical and peripheral details of the political process which can be readily manipulated by the quantitative methods of sampling, scaling, testing and content-analyzing—such matters as electoral statistics and mass media research ("who says what to whom through which channel"). Second, the behaviorist may take up his measuring rods and push on into the central areas of politics, ignoring their ambiguity and trivializing their content; in the words of Hans Morgenthau, he "can try to quantify phenomena which in their aspects relevant to political science are not susceptible to quantification, and by doing so obscure and distort what political science ought to know."[14] Finally, the behavioral scientist may abandon political realities altogether and retire to the heights of pure Method—with the vague intention of some day returning to the world when the master formulas have been computed and the tests for statistical significance are in.[15]

Implicit in the general retreat from politics which these attitudes reveal is a thinly veiled contempt for the political process as a whole. ("To Gary," reads the dedication of a volume on the psychology of politics, "in the hope that he will grow up in a society more interested in psychology than in politics.")[16] The sources of this hostility are at least twofold. It may result simply from the shock of recognition; when a behavioral scientist is brought to realize that the central concerns of politics do not lend themselves to his orderly and rational analysis, he is likely to conclude that they are, *ipso facto*, disorderly and irrational. Still more conducive to the breeding of contempt is the formal distinction which the behavioralist characteristically maintains between his role as scientist and his role as "citizen."[17] There is seldom any question in his mind as to which is the primary and superior role, judged in terms either of rationality or of responsibility. All of his scientific training, his accumulated knowledge and mature sensibility go into the construction of his professional image; what goes into his role as citizen is the seamier residue of human nature and experience: i.e., unconscious conditioning, unsublimated aggression and arbitrary preference. On one side the act of *investigation* (the process of research) is coherent, logical and eminently objective: in short, it is scientific. On the other side the act of *participation* (the process of politics) is incoherent, emotional and utterly capricious: in short, it is irrational. Hence a well-known psychologist of politics can confidently describe his relationship to his own research as follows:

> However much some of the attitudes studied may be anathema to me personally, such feelings are irrelevant and must be prevented from contaminating a purely factual and objective study. . . . For the same reason I have kept away from any suggestion as to the possible uses of the findings reported here. As a citizen, I have strong views on this point; as a scientist, I recognize that these views are value-judgments, and that they have no place in this account.[18]

If political beliefs and commitments must be quarantined and fumigated lest they contaminate the purity of the scientific enterprise—if, in effect, the scientist must check his civil obligations at the door of the laboratory—then the converse must also be true. When he leaves the laboratory for the world of affairs, he leaves behind him his professional training, his intellectual equipment and, in a word, his wits. He descends into an abyss of passion and prejudice, where all judgments are of equal weight (zero) and all choices merely arbitrary. He may possibly be permitted, as citizen, to act out his feelings and vent his repressed affect; but he must be careful not to assign any validity or value (other than therapeutic) to such role-playing. In this manner one of the most respected of sociological behaviorists, George A. Lundberg, concedes that "social scientists, like other people, often have strong feelings about religion, art, politics, and economics. That is, they have their likes and dislikes in these matters as they have in wine, women and song."[19] Social scientists, in other words, have every right to indulge the whims of their baser natures for liquor and legislation, courtesans and constitutions; difficulties arise only when it is supposed that the social scientist knows what he is doing in society —or the political scientist in politics.[20]

Lundberg: Science as Salvation

Lundberg has, as a matter of fact, gone a good deal farther than merely to assert that the activities of science and politics must be carefully distinguished—or even that the former is intrinsically higher and more honorable than the latter. In his vigorously affirmative reply to the question, "Can Science Save Us?", he leaves little room for doubt that one of the principal evils from which science can and should save us is politics itself. The revulsion from politics as a chaotic and "contaminated" pursuit, an attitude left implicit in the writings of most behavioralists, is here brought freely into the open and followed unflinchingly to its conclusion: which is that the scientific salvation of society consists in the elimination of politics altogether—that is, of contingency and conflict, contest and choice—in favor of the frankly authoritarian administration of public policies drawn up by social scientists and validated by scientific techniques.

Foremost among these techniques is the public-opinion poll, a technological discovery which, in the perspective of history, "may rank in importance with gunpowder, telephone, or radio," and through which at long last "can be resolved the principal impasse of our age, namely, the apparent irreconcilability of authoritarian control on one hand and the public will on the other."[21] (It might be observed in passing that this is, surely, an "impasse" only for authoritarian rulers seeking a plebiscitary semblance of legitimacy; for a democracy the problem is not how to become "reconciled" to authoritarian control but only how to avert and defeat it.) The extraordinary virtue of the public-opinion poll for Lundberg, as for numbers of the "pollsters" themselves, is that it promises to bridge this hypothetical impasse at a leap, by providing an infallible scientific substitute for the whole antique and clumsy apparatus of the democratic political process—a much "more delicate barometer of the people's will than is provided by all the technologically obsolete paraphernalia of traditional democratic processes."[22]

The substitution of science for politics would, moreover, go well beyond the elimination of obsolescent democratic machinery—and even beyond the implied renunciation of representative government in favor of a direct and instantaneous expression of "the people's will." It would replace the entire range of political choices and decisions with "scientific diagnoses," and render superfluous our primitive penchant for moral judgments and for such archaic "sentiments" as justice and virtue. Writing at the close of the second world war, Lundberg made plain the extent of his own political detachment in a singular passage:

First of all, the advancement of the social sciences would probably deprive us in a large measure of the luxury of indignation in which we now indulge ourselves as regards social events. This country, for example, has recently enjoyed a great emotional vapor-bath directed at certain European movements and leaders. Such indignation ministers to deep-seated, jungle-fed sentiments of justice, virtue, and a general feeling of the fitness of things, as compared with what a scientific diagnosis of the situation evokes.[23]

Along with this emancipation from the tyranny of morals goes the necessity of giving up a large vocabulary which has always accompanied that attitude. Among the old-fashioned concepts which must assertedly be abandoned or overhauled are "freedom, democracy, liberty, independence, aggression, discrimination, free speech, self-determination, and a multitude of others. . . ."[24]

But it is not only the "pious shibboleths" of our political vocabulary which the purgative of science will drive out.[25] It will require us also to abandon certain "deeply cherished ideologies" —*among them democracy itself*. "Scientists must recognize that democracy, for all its virtues, is only one of the possible types of organization under which men have lived and achieved civilization."[26] More importantly, democracy is not the only form of political organization under which *science* has flourished or can flourish. "The favorite cliché is that 'science can flourish only in freedom,'" says Lundberg. "It is a beautiful phrase, but unfortunately it flagrantly begs the question. *The question is, under what conditions will the kind of freedom science needs be provided?*"[27]

The question is, in other words, not under what conditions will the kind of freedom *men* need be provided; the essential freedom is the freedom of *science*, not that of humanity. "Political systems have changed, and they will change. Science has survived them all as an instrument which man may use under any organization for whatever ends he seeks."[28] But the instrument, as it turns out, will have something to say about its uses; for Lundberg goes on to call for a blanket delegation of authority under which social scientists, replacing the ward boss, will relieve the individual citizen of much of his obligation of citizenship.[29] Finally, he quotes with approval the pronouncement of a fellow-scientist that "the day is gone, and probably forever, when a successful state can base its politics upon the clamor of pressure groups or upon the uninformed beliefs of the majority, even measured numerically by tens of millions."[30]

With this conclusion, Lundberg's thesis has plainly turned upon itself and become indistinguishable from its antithesis. Social science, which was to be at the disposal of society, in the end disposes of society. It is the word of the scientists, not the "voice of the people," which is to prevail after all.[31] This denouement, however much it violates the author's posture of disinterest, follows logically enough from his semiarticulate major premise: that is, his contempt for the politics of democracy and the broad range of human strivings and experience which go to make it up.

But, despite all this, it should not be thought that Lundberg's hypothetical elite of scientists wishes to impose upon the world any substantive political philosophy or program. The paradox of this peculiar movement is that, even while it presses for the prerogatives of power, it is still fundamentally "value-neutral." Its interest in politics is only technical and administrative; its rationale is the bureaucratic ethos and the gospel of efficiency. The one "goal-value" which it seeks to actualize is a nebulous climate in which the scientist in all his forms (as researcher, technician or administrator) may function unmolested. Given this minimal condition of political immunity, the social scientist as portrayed by Lundberg has no further preference and no deeper concern. Indeed, the supreme ideal of social science, in this view, is to attain to so celestial a degree of detachment from worldly affairs as to be instantly recognized by all political persuasions—Fascist, Communist, democratic, or whatever—as at once perfectly harmless and perfectly available. Lundberg's statement of this thesis of neutrality is sufficiently remarkable to be worth quoting at some length:

> Physical scientists are, as a class, less likely to be disturbed than social scientists when a political upheaval comes along, because the work of the former is recognized as of equal importance under any regime. Social science should strive for a similar position. . . . If social scientists possessed an equally demonstrably relevant body of knowledge and technique of finding answers to

questions, that knowledge would be equally above the reach of political upheaval. *The services of real social scientists would be as indispensable to Fascists as to Communists and Democrats, just as are the services of physicists and physicians.*[32]

And finally: "There is nothing, then, in the proposals of this book or in the social order which it envisions which cannot be supported by the adherents of every religion, every political and every economic faction interested in the efficient and economical achievement of whatever objectives they pursue."[33]

It would be difficult indeed to improve upon the clarity of this declaration. It presents to us with candor and conviction the vision of a social order in which the goal-values (whatever they may be) of the governing authority (whatever he or it may be) will unhesitatingly be translated by a disciplined corps of social technicians into action programs of cultural and behavioral engineering. We have encountered this vision before; but what once may have seemed only the hallucination of morbid minds, or at best the fantasy of science fiction, has begun to take on the dimensions of genuine plausibility. Indeed, the technical means of its fulfillment seem almost within our grasp. In the efficient and economical implementation of the "given" social goals—in (for example) the perfection of the decision-making process and its consensual validation by means of subliminal persuasion; in the rationalization of intergroup conflict and the elimination of deviant behavior; in the mounting accuracy of mass-behavioral prediction, and the approaching uniformity of everybody's conditioned responses—in such practical matters are foreshadowed the ultimate triumph of that "positive spirit" first adumbrated by Comte and the Saint-Simonians, along with the vindication of the neglected prophet Watson—and the last laugh of the founder of Walden Two.

The Values of Neutralism

It is a well-known comment that the abdication by value-neutral social scientists of any responsibility for the use or consequences of their research has tended to leave them, in fact as well as in principle, available "for hire."[34] By their own express renunciation the goal-values, the ends and purposes for which they are solicited, are simply "given." (Theirs not to reason why, as someone has surely said; theirs but to quantify.) The actual result of this condition of voluntary servitude is a matter of record: behavioral science is thereby placed effectively at the service—not of "society," as the favored expression goes—but of the corporate interests (private and quasi-public) which represent the highest bidders in the academic marketplace. The various exploitive and manipulative implications of this interested patronage have been well-described elsewhere, and need not be recapitulated here.[35] But there is one significant aspect of this phenomenon which has escaped general notice: that is, the curious contradiction in which the attitude of ethical neutrality, if consistently adhered to, involves its advocates.

It is Professor Lundberg, once more, who has most clearly illuminated the underlying confusions of value-freedom. His views have been set forth in the course of a discussion of Gunnar Myrdal's monumental study of the Negro in the United States, *An American Dilemma*.[36] It hardly needs to be remarked, as a preliminary, that Myrdal's investigation was systematically organized from start to finish around a set of moral and political value-judgments ("the American Creed") which formed the fundamental frame of reference for the massive research making up his two-volume enterprise. Far from taking exception to this value-drenched approach, Lundberg has surprisingly approved of it as entirely consistent with his own neutralist precepts: "The scientist may elect, either from personal interest or because he is paid [sic], to appraise the situation from a particular viewpoint. No one suggests that such a selection of one's problem represents a departure from strict scientific work or that mysterious problems of 'value' are involved."[37]

This statement is noteworthy on at least two counts. First, it is paradoxical at best to maintain that the selection and appraisal of a problem-situation from "a particular viewpoint" raises no question of values or of departure from strict (value-neutral) scientific work. Even if the point were arguable in principle, however, in this case it runs squarely up against Myrdal's outspoken affirmation of his own pervasive value-orientation—as contained, most notably, in the famous "Methodological Note" appended to his study.[38]

Why has Lundberg felt impelled to swallow such a camel? The explanation would seem to lie in the first sentence of the quotation directly above. It is clear that the consistent value-neutralist deems himself in no position to question the nature or source of the goal-values governing any research assignment to which he is called. But if all goals and viewpoints are equally permissible, and if anyone may stipulate them, it follows that "the scientist [himself] may elect . . . to appraise the situation from [his own] particular viewpoint." The interesting possibility thus arises of a situation in which the strict value-neutralist may freely prescribe the values of his own research and proceed to appraise his data from their standpoint—exactly like anyone else. At this juncture it does not seem excessive to suggest that the principle of value-neutralism has lost its meaning.

But this does not yet fully explain the puzzle of Lundberg's acquiescence in the injection of value-assumptions by the social scientist "either from personal preference *or* because he is paid." It is possible that what the author most intends to justify is the latter contingency—one which is more evidently congenial with the monastic vows and alms-seeking propensities of the neutralist order. However, it is here that the most persistent of ethical questions arises. Can the social scientist serve two masters—Truth and the Corporative Sponsor? When he has rendered unto the corporation that which is the corporation's, how much is left to the idols of the laboratory? Is the scientist's allegiance to truth only an obligation to obey the ground rules and observe the amenities of something called Scientific Method—selectively perceiving all the trees within his field of vision, while selectively overlooking the forest?

Even on the behaviorist's own rigorous terms, it may be

questioned whether the deliberate exclusion of the whole context of conditions and consequences—the matrix of social and political experience from which his data are abstracted and within which they find their meaning and effect—is not in the end an avoidance of all claim to relevance and significance. Is it good science (let alone good sense) to call a halt before the logical end-point of inquiry, without drawing the full conclusion? Or, to put it the other way around, is it good science to ignore the origins and assumptions, the starting-point and reason-for-being, of one's investigation? How free is the spirit of inquiry when confronted with an iron curtain at both ends? And what is there left to social science by this self-denying ordinance other than the clerical performance of checking and tabulating, classifying and accounting?[39]

Given the behavioralist's posture of indifference toward "subjective" values and qualitative choices, this emasculated image of science would seem to be inescapable. There is clearly no place in it for the fretful sleepwalking and creative discontent that are commonly thought to precede scientific discovery—at least by the great discoverers—nor for the "heuristic passion" and warmth of commitment which others have believed to accompany and enrich its career.[40]

It is notable that the bleak definition of scientific method advanced by Lundberg as the basis for a "natural science of society" has been singled out by a distinguished natural scientist as a model of superficial inaccuracy and misunderstanding.[41] That formulation, according to James B. Conant,

> is a typical description of what is often called scientific behavior, but I venture to suggest it is not a description of the characteristic way the natural sciences have advanced; it is rather an account of the use of very limited working hypotheses not dissimilar to those employed in everyday life.[42]

What Conant means by the limited hypotheses of everyday life is made clear by another of his observations on the general attitude exemplified by Lundberg: "The activities I have listed are not science; they are either exercises in logic, or the repetition of activities once significant in the advance of science, or essentially trivial and tiresome mental operations for some practical end, entirely equivalent to making change."[43]

It is a commonplace of the sociology of science (although perhaps not yet of the sociology of sociology) that the assumptions and beliefs of scientists are not apart from, but a part of, their culture and society.[44] The effort to be "neutral," in social science as in power politics, generally has the consequence not of rising above the battle but of being caught in its crossfire. In fact the posture of value-neutralism, far from being the unique prerogative of behavioral scientists, is a prominent and pervasive characteristic of the very *milieux* toward which such students have been gravitating. It should surprise no one to discover that, as behavioral scientists come more and more to resemble Berelson's "teams of technicians" and less and less to resemble independent scholars, they should also begin to take on other attributes typically associated with the technician. Indeed, considerable light may be thrown upon the cult of neutrality in social science by viewing it as part of a much broader movement—that which Max Lerner has described as "the Great Withdrawal of the Neutral Technicians." For Lerner, as for other observers, one of the most ominous developments of our time is "the central neutrality of the technician (whether engineer, manager, or executive), in the sense of his dissociation from passion, commitment, or value other than his own skill in execution."[45] As this comment suggests, a study of the features of the neutral technician in various sectors of American society discloses a striking resemblance to the profile of the value-free behavioral scientist. "The role of the Neutral Technician," Lerner has concluded, "thus casts its shadow over the whole present era. It becomes the Great Withdrawal, or—as Erich Kahler has put it—a kind of nihilism of values along with an exaltation of techniques."[46]

It is therefore to be expected that, as academicians become technicians, they should also come to share the special ethic (the "bureaucratic ethos," in C. Wright Mills's term) of the technical fraternity. To be sure, the mandate of neutrality has differing connotations for different groups of technicians. For the executive of industry it is bound up with a "marketing orientation"[47] contributing to his occupational mobility and flexibility; for the military it is a formal tradition imposed from above, alternately worn with pride and irritation; for the Administrative Class of the government establishment it is a vital ingredient of guild spirit—although one not necessarily incompatible with an invisible hand in policy-making. It is pre-eminently the latter group of technicians, the governmental bureaucracy, whose well-articulated attitudes toward neutrality parallel those of the neutral behavioralists. Indeed, at a certain point the two sets of attitudes cease to run parallel and become overlapping and interwoven. It should thus be useful to our inquiry to take a closer look at the sources and implications of the older tradition of neutrality in public administration.

The "quest for neutral competence," as Herbert Kaufman has demonstrated, began in American government about a century ago and has continued as a central tendency (although not the only one) down to the present day. The primary value of this search was ability to carry on the work of government expertly, according to objective standards rather than personal, political or moral obligations. "The slogan of the neutral competence school became, 'Take administration out of politics.' "[48] The desire for emancipation of the science of administration from the lively art of politics reflected both a veneration of objective scientific methods and a distaste for politics resembling that which we have traced in the writings of some behavioral scientists. In the words of Dwight M. Waldo:

> To declare independence from politics was almost a necessity in making the claim to science plausible, for the rough and tumble of politics seemed completely at odds with the order of the laboratory that connoted science. Or put the other way around, the cool, calculating, rational spirit of science seemed to demand divorce from the passion and chance of the political realm and its seeming disorder.[49]

For the most part, to be sure, the early administrative neutralists couched their arguments in terms broadly congenial with democracy, on the unchallenged assumption that the technical rationalization of the public enterprise through objective administration would promote the cause of democratic leadership and representative government. Some proponents of neutral competence, however, were even then explicit in their antagonism to the disarray of democratic politics; and indeed the origins of the movement, as Kaufman has shown, betray an unmistakable elitist bias. "The disillusionment of some was so thorough," he writes, "that they lost faith completely in representativeness, in the capacity of a people to rule themselves, and returned to advocacy of rule by an aristocracy of talent."[50]

The idea of "neutral competence"—of a hierarchy of administrative technicians above the political battle if not beyond all good and evil—is today as strong as ever.[51] But it is no longer uncontested or even quite predominant in its own field. The old dichotomy of politics and administration has in fact been widely abandoned by administrative specialists both in and out of government in favor of something like a tacit reunion of the two.[52] This seemingly startling reversal of form is not, however, quite so revolutionary as it appears. For whatever may have been the formal conception, the divorcement of politics and administration was never really meant to leave them separate but equal: it rather implied, insofar as its will could be made effective, the *subordination* of politics to administration in the general process of government—if it did not also imply, in the fullness of time, the banishment of politicians from the temple. But this inversion of the two functions could only mean, as in practice it did come to mean, the covert entrance of administrators into politics. If for no other reason than to safeguard their preserve from alien policy-makers and other political poachers, the neutral bureaucrats found themselves forced to take the initiative in policy matters and to secure their own alliances with outside powers—in short, to play politics. "The components of the 'neutral' bureaucracy, by virtue of their expertness and information and alliances, have become independent sources of decision-making power. . . ."[53]

The faith in neutral competence, for some at least among

its advocates, was fundamentally a belief in the superior competence of the neutrals: in the right of a saving remnant of experts to control the government, and so to restore to rational order the deplorable confusion which democratic politics had left in its wake. At bottom, these "neutralists" were never really neutral; they were motivated by a profound repugnance to the politics of democracy which led them to seek not only a new and balanced alignment of forces but a virtual reversal of the traditional relationship of politics and administration. The most ambitious of the reformers were not content with the replacement of the spoils system by the civil service; they looked forward, like their latter-day counterparts within the social sciences, to nothing less than the replacement of all political contingency and chance by the scientific certainties of neutral competence. Nor has their quest for the utopia of the Administrative State yet been abandoned. Today hardly less than in the past, to quote Kaufman once more,

> it might be inferred that some reformers distrust all politicians and electorates and pin their hopes on the expertise and efficiency of a professionalized bureaucracy. They seem to be moved not merely by a concern for governmental structure but by political values that include an implicit contempt for what we ordinarily understand to be the democratic process and an explicit respect for an aristocracy of talent that borders on a latter-day faith in technocracy.[54]

It is not quite accurate to speak (as above) of the scholarly "counterparts" of these administrative technicians. For although the *roles* of scholar and administrator remain distinct,[55] they have frequently (since the days of the New Deal and more particularly since the second world war) been occupied by the same person. Moreover, the vast professional literature of public administration, including the dispensation of its "principles" and values, is mainly the work of academic specialists. The creed of neutral competence may have had its origin outside the academy; but it has since drawn powerful support from within its walls. Most instrumental of all in reinforcing and revitalizing the neutralist creed have been the writings of the modern school of political and administrative behaviorists, led most prominently, perhaps, by Harold D. Lass-well.[56]

It is the distinctive claim of this self-consciously scientific school that the student of political behavior can, in simple truth, have his value-cake and eat it too—that his behavioral science has somehow contrived to be *both* pure and applied, value-purged and value-urged, at once indifferent and dedicated to the democratic cause. The divorce of policy (value) and administration (neutrality) is, it now appears, only a formal separation, and the pair still live together in virtue: for the administrative science of behaviorism is in truth a "policy science."[57]

The links in the chain of reasoning which support this improbable conclusion may be readily identified. In the beginning was the theorem; what is assumed as a postulate is that the "method" of natural science can (indeed, must) be brought to bear with utter fidelity upon the study of human affairs—if we are to be saved from ourselves and from our political tradition. The philosophy of politics thus emerges as a branch of the "philosophy of science"—a term which (whatever scientists may think or say about it) possesses for its advocates a highly specific and delimited meaning, as positive as it is logical.[58] It is positive most of all about what it does *not* mean: among other things, it does not mean indulgence in broad-gauged speculation concerning what are known elsewhere as the Great Issues;[59] rather, the treatment which these issues deserve and get is primarily that of linguistic analysis, which permits any discussion of them to be conveniently allocated into either of two bins: one labeled "fact-statements" (or scientific propositions), and the other labeled "value-judgments" (or assertions of preference). Sentences of the first kind possess "truth-value," being statements about matters of fact which may be empirically verified; sentences of the second kind possess no truth-value, being not statements in the strict sense at all but emotive utterances confined to the expression of subjective feelings.[60]

It may be surmised that, on these black-and-white terms, the considerable body of speculation and discussion on political subjects which has accumulated over the past twenty-five hundred years barely suffices to cover the floor of the first of these bins—while it abundantly overflows the second receptacle—which then provides the basis for a research file on the folklore and mythology of politics. For it would be a mistake to think that, in this positive science of politics, the second of the two bins is regarded simply as an ashcan. To be sure, at the hands of some intrepid spokesmen all judgments of value and works on political philosophy are either burned as trash or dispatched through intercampus mail to the departments of arts and letters.[61] But this is no longer characteristically the attitude of political and administrative behaviorists. All the "values" which men have cherished, and their metaphysical celebrations in the annals of the academies, are now regarded as important (if cryptic) clues to the analysis of behavior—an essential part of the objective data of human experience which it is the prerogative of the social sciences to assemble and assimilate (or, more accurately, to collate and compute).[62]

Politics and its values, then, are still held off at arm's length —but only in order to draw a bead on them. "In this view," as Professor Waldo has noted of political positivism, "values (as reflected in human behavior) are to be studied empirically, from the outside, just as any other phenomena are studied by a scientist."[63] They are studied, however, not for their own sake or on their own terms—as substantial propositions or statements of fact to be confirmed or disconfirmed—but rather as nonrational and hence irresponsible indices of the authentic (i.e., authenticatable) meanings which lie somewhere behind them.[64] The semantics of analytic philosophy thus joins the mechanics of psychological behaviorism in the denial that cognitive meaning (or honest claim to truth) is to be sought or found in the verbal gestures of "conation" through which men communicate their private sense of what is valuable.[65] In not dissimilar fashion, this scientific philosophy of politics joins hands with the more venerable conviction of "scientific socialism" that all such affirmations (except perhaps one) are to be understood as ideological smoke screens shielding a subterranean deposit of opportune interests.[66] The viewpoint, finally, has another instructive parallel in the analytic method of Freudian psychology, whose primary concern with the language of profession and confession is with its latent (repressed) rather than with its manifest (rationalized) content.[67]

All of these diverse approaches to the study of behavior are alike in their reductive attitude toward "values," and hence in their purgative outlook toward the language of politics. It is notable that they also converge in the development of the highly individual but profoundly influential perspective of Harold Lasswell.[68] In his prolific writings on political subjects over more than a generation, these several approaches figure recurrently as alternate and complementary themes—held together, however precariously, by an underlying devotion to the spirit and cause of objective science.

Lasswell: The Politics of Manipulation

Lasswell's debt to Marxism, possibly his first in point of time, is declared in various places—most directly perhaps in his attraction to its "methodology,"[69] less directly in the elaboration of his theory of political and social elites. It is a revealing index of his neutrality as scientist that Lasswell could attempt to divorce the method of Marxism from its goal-values and to contemplate its plausibility as the means toward a total solution of the "World Revolutionary situation of our times."[70] For it was a "total" solution to world insecurity in a time of world revolution that Lasswell was systematically seeking in his major writings of the 1930's—a solution total alike in its scope (world-wide) and in its method.

> The emphasis which is here put upon the importance of appraising the total meaning of the developing situation for social values is in many respects parallel to the viewpoint introduced by Marx and Engels into modern social theory. . . . their perspective was political-totalistic, for they sought to assess the meaning of every detail of the total situation for preserving or demolishing particular value pyramids.[71]

As this makes clear, total solutions to the problems of politics, if not of all human experience, are made possible by our increasing capacity for "preserving or demolishing particular value-pyramids"—specifically through the skills of conversion, manipulation and persuasion. The classification of these skills forms the subject matter of political science; in fact, it would seem that this activity is the only function of political science. The study of politics is simply the study of influence and the influential: "Those who get the most are *elite;* the rest are *mass.*"[72] "The elite preserves its ascendancy by manipulating symbols, controlling supplies, and applying violence. Less formally expressed, politics is the study of *who gets what, when, and how.*"[73] Such a study is, of course, strictly neutral and value-free; where the philosophy of politics "justifies preferences," the science of politics merely states conditions.[74]

But it is not quite that simple. The study of politics is both "the science and *art* of *management*";[75] it contains not only a "contemplative" component but a "manipulative approach" as well—one which views events in order to discover ways and means of gaining goals, and which brings "the attitude of the analyst much closer to that of the agitator-organizer."[76] Neutral contemplation, then, is only one side of the coin of scholarship. Like the yogi of another contemplative tradition, the student of politics is obliged to go down from the mountain into the marketplace and put his art and science to work. "Only vitalizing sentiment can transform the terms of a contemplative analysis into the slogans of social movements and into the practices of established social institutions. The re-direction of culture requires skill in propaganda, in violence, in organization; skill in analysis is not enough."[77]

There is, besides this, a more immediate urgency. His contemplative analysis of "skill politics and skill revolution" has disclosed to Lasswell that in our own time the most potent of all skills is that of propaganda, of symbolic manipulation and myth-making—and hence that the dominant elite must be the one which possesses or can capture this skill. In fact the overriding problem of world unity, of a universal solution to insecurity and conflict, is that of devising a "nonrational world myth" capable of commanding the allegiance of men everywhere. Lasswell's description of the power elite and its universal myth reads like nothing so much as an academic paraphrase of the Grand Inquisitor, with his famous *miranda* of "miracle, mystery and authority."[78] But who is to get this ultimate power, when, and how? An obvious contender might seem to be the "politician," the elective agent in a democratic polity. However, Lasswell reviews the credentials of this "personality type" in a manner markedly less contemplative than contemptuous:

> Whatever the special form of political expression, the common trait of the political personality type is emphatic demand for deference. When such a motive is associated with skill in manipulation, and with timely circumstances, an effective politician is the result. The fully developed political type works out his destiny in the world of public objects in the name of public good. He displaces private motives on public objects in the name of collective advantage.[79]

Clearly the politician, democratic or other, is not what is wanted (not even by analysts who state no preferences). Nor is it the dictator, the oligarch, the hereditary monarch. One skill-group alone within the "skill-commonwealth" stands out as a logical choice. It is that group which is the most skilled in the methods of verbal manipulation and myth-making—namely, the academic "symbol specialists" or social scientists. In fact, for Lasswell the manipulation of symbols (in however exotic or detached a form) *implies* the manipulation of men; which is to say that nearly all who write are implicit propagandists. "Skill in handling persons by means of significant symbols involves the use of such media as the oration, the polemical article, the news story, the legal brief, the theological argument, the novel with a purpose, and the philosophical system."[80]

This quasi-Marxist indictment of the intellectual as apologist and lackey is, no doubt, widely deserved and reasonably relevant so long as its cynicism is qualified by the recognition that even scholars may upon occasion seek the truth. (If they are "nothing but" lackeys, the argument simply nullifies itself, as critics of the "unmaskers" from Marx to Mannheim have pointed out; it then remains only to unmask the interests for whom the unmasker is himself masquerading.)[81] Lasswell makes no such qualification, but he does soften his indictment in a curious way: if the symbol-specialists must always be found in the service of some interest (or "myth"), that interest may at least be *self*-interest and the myth one of their own creation. "The pattern for mythmaking by intellectuals," he writes, "was set for our society by Plato, who dreamed poetically of the 'philosopher king' in whom omniscience was at one with omnipotence."[82]

With this suggestion the marriage of the contemplative and manipulative components of political science is consummated—through the strange bedfellowship of Machiavelli and Plato. With respect at least to this favored elite, Lasswell is no longer dealing so much with ideology as with utopia. In the face of the saving remnant, his mask of indifference slips noticeably askew. Thus, contemplating the "possible though no doubt unforeseeable outcome of academic activity" which seeks to "rearrange the value pyramids for the benefit of the specialists," Lasswell makes this candid confession of interest:

> It is indisputable that the world could be unified if enough people were impressed by this (or by any other) elite. The hope of the professors of social science, if not of the world, lies in the competitive strength of an elite based on vocabulary, footnotes, questionnaires, and conditioned responses [sic], against an elite based on vocabulary, poison gas, property, and family prestige.[83]

Once again, the hidden value-premise of neutralism has asserted itself; the commitment of the social scientist to the *contemplation* of behavior is at the same time an agreement to its *manipulation*. The elitist conclusion has all along been implicit in the premises; once the vocation of the scientist is defined as a "skill," and his role as that of a symbol "specialist," we have passed from the realm of disinterested science into that of technology—more particularly, of social and behavioral engineering. The "activist" nature of Lasswell's applied political science is consistently betrayed by a vocabulary so militant in metaphor as to seem almost military in its derivation. Not merely are ideas weapons; philosophies are strategies, analysis is a kind of reconnaissance, and specialist-elites resemble nothing so much as cadres. (The following titles in a collection of Lasswell's essays, *The Analysis of Political Behavior*, are representative: "Skill Politics and Skill Revolution," "The Garrison State and Specialists on Violence," "Policy and the Intelligence Function: Ideological Intelligence.") What is most vividly reflected by these urgent metaphors is the author's underlying sense of embattlement and crisis, of a developing "world revolution" with apocalyptic overtones—suggesting a universal insecurity which demands drastic therapies and total solutions, if not indeed the transformation of all human culture and politics.

It is therefore not surprising to discover that, along with his preoccupation with power and elites, Lasswell was continuously concerned during the 1930's with the application to politics of the therapeutic insights of psychoanalysis.[84] He had prepared himself for this task by training and study under Freud himself; but, as with some other of his adaptations, Lasswell took for granted the political relevance and propriety of the "goal-values" of Freudian psychology. The goal of politics became literally the goal of the clinic: that is, the "obviation of conflict." In his *Psychopathology and Politics*, he laid the foundations and announced the objectives of an ambitious "politics of prevention":

> The problem of politics is less to solve conflicts than to prevent them; less to serve as a safety valve for social protest than to apply social energy to the abolition of recurrent sources of strain in society.
>
> This redefinition of the problem of politics may be called the idea of preventive politics. The politics of prevention draws attention squarely to the central problem of reducing the level of strain and maladaptation in society.[85]

As this and other of Lasswell's writings suggest, the ultimate objective of the politics of prevention, for a democracy at least, would seem to be the *prevention of politics*. For the urge toward an "obviation" of political conflict embodies a plain distrust of the processes of peaceful conflict which democracy has encouraged and institutionalized—distrust, not because they are inadequate to evoke a genuinely free expression of the interests of differing individuals and groups, but precisely because they do such a good job of it. Lasswell asserted that the time had come to abandon the traditional assumption that "the problem of politics is the problem of promoting discussion among all the interests concerned in a given problem." His Freudian insight revealed instead that "discussion frequently complicates social difficulties, for the discussion by far-flung interests arouses a psychology of conflict which produces obstructive, fictitious, and irrelevant values."[86]

Underlying this failure of the democratic dialogue is a more fundamental malaise: the failure of democratic man. The findings of personality research, we are told, "show that the individual is a poor judge of his own interest. The individual who chooses a political policy as a symbol of his wants is usually trying to relieve his own disorders by irrelevant palliatives."[87] In the light of this corrosive analysis of the decision-making citizen, it can only be concluded that "serious doubt" is cast upon the efficacy of public discussion as a means of handling social problems.

In short, according to the psychopathology of politics, the presumption in any individual case must be that political action is maladjustive, political participation is irrational, and political expression is irrelevant. To be sure, the citizen may find a temporary relief from tension through "the displacement of his private affects upon public objects. *But the permanent removal of the tensions of the personality may depend upon the reconstruction of the individual's view of the world, and not upon belligerent crusades to change the world.*"[88] The acceptance of Freud carries with it, in the end, the repudiation of Marx.

The objectives of the science of preventive politics are, then, both to remove the internal causes of "faction" (or at least keep them safely repressed)[89] and to eliminate their external effects (i.e., discussion and debate). Lasswell seems to have recognized that such a politics could hardly be called democratic, that it was more likely to suggest authoritarian restraint. For he submitted a brisk disclaimer of any preference for dictatorship, but at the same time showed himself oddly impatient with those who would venture to raise such questions: "Our thinking has too long been misled by the threadbare terminology of democracy versus dictatorship, of democracy versus aristocracy. Our problem is to be ruled by the truth about the conditions of harmonious social relations."[90] Unfortunately the "truth," being disembodied, can rule only through some human agency; which is to say, as Lasswell thereupon did say, that we ought to be ruled by those best equipped to discern this truth and to administer these harmonious relations—namely, "the social administrator and social scientist."[91]

Lasswell's therapeutic politics of prevention thus provided a congenial corollary to his politics of manipulation. It reinforced the preference for an elite of social scientist–administrators dedicated (in some manner they were prohibited from understanding) to a vision of scientific truth and social harmony. ("In some vague way," as he put it, "the problem of politics is the advancement of the good life. . . .")[92]

But there was also a *logical* as well as a psychological aspect to his developing perspective. Just as the conflict and confusion of political activity might be resolved through the process of "purging" the citizen, so the ambiguity of political discourse might be resolved through the process of purging the language. Underlying all the separate interests and tangents of Lasswell's career, as noted earlier, has been his profound dedication to the cause of "science." In his hands the application of science to politics—whether through objective analysis and description, through psychotherapeutic diagnosis, or through the linguistic approach of "quantitative semantics" (content analysis)—has always meant a systematic effort toward clarification and simplification of the subject matter. By far the most rigorous and painstaking of the scientific techniques brought to bear by Lasswell has been his analysis of the language of politics (set down, for the most part, in books and articles of the 1940's).[93] In one sense, at least, this semantic contribution has also been his most fruitful: it has stimulated an ever-growing number of research undertakings aimed at the statistical treatment of political discourse;[94] and, perhaps more significantly if not more fruitfully, it has been instrumental in introducing into political science the viewpoint and assumptions of that contemporary philosophical persuasion which has taken upon itself the title of "philosophy of science"—but which is better known (if not better understood) as logical positivism.[95]

Like the quantitative semantics to which it bears a family relationship, the characteristic intention of logical positivism in any of its applications is to clarify, simplify, objectify, and, in short, to set matters straight.[96] As we have noted before, its approach to a given field of investigation is essentially to attack it through its "discourse"—which means in effect to isolate language from subject matter by scanning documents, not primarily for substantive content but for logical coherence.[97] Armed with the proper semantic yardsticks, the modern positivist may move with confidence through any thorny field of science or social science, even of "humane letters,"[98] without the necessity of prior briefing or professional acquaintance with the terrain—readily separating, as he goes, sense from nonsense, truth from metaphysics, reality from myth, and (in a methodological sense if no more) right from wrong.

This is the purport and procedure, for example, of the remarkable venture into "political logical positivism"[99] conducted by Lasswell in collaboration with an authoritative spokesman of the scientific philosophy, Abraham Kaplan. At the outset of their book, *Power and Society*, the two authors are at pains to establish the scientific standpoint of their inquiry and to distinguish it from the normative exercises of traditional political thought. The work is to be regarded as an attempt to formulate the basic concepts and hypotheses of political science. "It contains no elaborations of political doctrine, of what the state and society ought to be."[100]

Yet somehow, for all its abstention from political doctrine, their book turns out to be "in many ways in accord with the grand tradition of political thought."[101] Not only does it possess "continuity . . . with the major currents of the political thought of both past and present," but it even "conforms . . . to the philosophical tradition in which *politics and ethics have always been closely associated*."[102] Having been carried thus far, the reader should not be astonished to learn that the whole enterprise of *Power and*

Society is properly to be understood as conditioned by, subordinated to, and thoroughly baptized in the goal-values of the "democratic ideal."[103] Once again the familiar term "policy science" makes a ceremonial appearance, as a likely corral for the mutual husbandry of fact and value; moreover, the authors crown their labors at semantic clarity with the invention of a still more disarming synonym for the social studies: i.e., the *"value sciences."*[104]

These strenuous exertions both to have and eat the cake of customary value present, as plainly as we have yet observed it, the perennial agony of scientism: the dilemma of the confirmed neutralist upon whose ethical indifference righteousness keeps breaking in. Lasswell and Kaplan are capable of declaring without compunction that "as a citizen, a moral person, the scientist has his own preferences, goals, values"—and even that "all his acts, *including his acts of scientific inquiry*, are subject to self-discipline by moral aims."[105] Their own self-disciplining aims are "those of the citizen of a free society that aspires toward freedom"; accordingly, special attention must be given to the moral prerequisites and implications of a free society, and in fact the entire study of *Power and Society* is expressly undertaken from that standpoint. It can even be affirmed (with no consciousness of redundancy) that such moral considerations, "to the degree that they are *warranted* and *fruitful*," will serve to "foster freedom."[106]

Up to this point, the most value-committed of old-fashioned moralists could find little to complain of; the drift of the book is unmistakably in the direction of *having* one's values. But then, in a sudden gesture, they are eaten: "But we are not concerned with the justification of democratic values, their derivation from some metaphysical or moral base. This is the province of political doctrine, not political science."[107] The ruling values are to be celebrated but not examined.

Throughout his writings of the past two decades, Lasswell has repeatedly postulated "human dignity" as the ultimate value "to be realized in theory and fact"—or, at a minimum, as the *primus inter pares* of the goal-values felt to sustain democratic society.[108] Yet never, in all those years, has this elusive absolute of human dignity itself been subject to examination in the light of philosophic precedent and the history of moral discussion; it is merely "postulated," and the scientist adjured to get on with the serious business of implementing it in theory and fact. The particular form of implementation which Lasswell has come to prefer is that of a special science within the policy sciences called, simply, the "science of democracy"—which is "restricted to the *understanding* and possible control of the factors upon which democracy depends."[109] The "understanding" of the factors upon which democracy depends does not extend, as we have seen, to the comprehension of its fundamental postulates, such as that of "human dignity." Indeed, the factors which are to be understood and controlled, although regarded as "democratic values," are of a substantially different order. "By the term 'value,'" Lasswell says, "we refer to a category of 'preferred events.' . . . At present the following eight terms appear to provide a workable list: *power, respect, affection, rectitude, well-being, wealth, skill, enlightenment.*"[110] Equipped with "some such list," we are in a position to consider any community "according to the old formula: Who gets what (values) when and how? If we think of democracy as general shaping and sharing, despotisms are at the other end of the scale, characterized by the concentration of values in relatively few hands."[111]

The most striking thing about this workable list of democratic values is that they are in no discernible sense democratic; they are simply the "preferred events" of men everywhere. To say, then, that the difference between democracy and authoritarianism is one of general, rather than exclusive, "shaping and sharing" of such universal prizes is to imply that democracy has no values distinctively its own but only a somewhat broader spread of power, skill, wealth, esteem, and so on. In this spirit it might plausibly be argued that if Hitler's Germany had only opened its avenues of mobility and co-optation a little wider, it would have become significantly more democratic. On these terms, again, there is a substantial "quantity" of democracy within the Soviet Union today, by virtue of its relative equality of social opportunity, its assiduous endorsement of skill, its espousal of well-being, its keen devotion to "enlightenment," and its pronounced affection for affection.

What has gone wrong? The source of this confusion over the nature and order of political values would seem to lie, in part at least, in the remarkable innocence with which an eminent political semanticist, for all his preoccupation with linguistic clarity and conceptual rigor, embraces the most nebulous and recondite of moral norms: that of "human dignity." All of the elaborate and exacting paraphernalia of modern scientific research is to be placed at the service of a concept so mutilated by rhetorical abuse as to be not merely dubious but actively dangerous until it has received at least the minimum degree of attention and specification which derives from what Hocking has called "ethical common sense."[112] But such specification Lasswell, by what seems an opportune oversight, does not provide.

Since he is not concerned to know the heritage and character of his ultimate goal-value, there is no evident uneasiness in Lasswell's persistent conjunction of "human dignity" with "manipulation," as carried out by the techno-sciences of political prevention and behavioral reform. (In an early statement he observed: "It may be that the manipulation of collective opinion for the sake of raising the prestige of science will contribute towards this sense of unity of man with man.")[113] The assumption which alone would seem to justify this linkage of contrarieties is that the "dignity of man" is not (as others have supposed) an inherent attribute of his humanity, nor the civilized expression of a categorical imperative, but a strategic objective to be achieved in some rational future. Viewed in these terms, there is nothing inconsistent in Lasswell's formula; if dignity is an objective goal, a *policy* to be implemented in theory and fact, then engineering may well provide the means, and men may plan for the realization of dignity as now they plan for slum clearance. As Lasswell has stated, "If our moral intention is to realize a democratic society, we need a science of democracy to implement the goal."[114]

Once more, as so often in our survey, the vision of a "techno-scientific culture"—a future democracy shaped by social scientists and shared by a homogenized public—comes into focus. "The forecast remains," writes Lasswell; "the world is moving toward homogeneous social structure, regardless of whether political unipolarity is early or late. . . . A new level of techno-scientific culture must be shared widely before its full benefits can be attained."[115]

On its own assumptions, this progressive concern for the expansion of benefits and the resolution of conflicts is doubtless unexceptionable. But alternative assumptions are available. Once human dignity is regarded not as a future by-product of social engineering but as an inherent quality of man *qua* man—more to be safeguarded from external encroachment than "implemented" by external fiat—the prospect of a manipulated dignity becomes less attractive. The unwitting inversion of values to which the policy science of democracy points is concretely illustrated by many of the concepts and programs of present-day public welfare (themselves the product of convergent traditions, such as those of Freud and Marx, not unlike the basic ingredients of Lasswell's thought). As in his case, the enhancement of human dignity and personal freedom is an avowed objective of social welfare; nevertheless, all too often, it is the dignity and freedom of the person as client which are oppressed and jeopardized by the manipulative propensities of welfare programs and programmers.[116]

In the elaboration of his policy science and its filial "science of democracy," Lasswell has not abandoned his earlier contemplative-manipulative science—his handbook for the prince. What he has done, with perfect good will, is to place it unedited at the service of democracy—just as, before him, John B. Watson had placed his system of behavioral conditioning and control at the disposal of society. There is no question here of the integrity and earnestness of his moral concern; all that is in question is whether the techno-scientific skills of mass manipulation and mythopoeism, of political therapy and behavioral engineering, can be said to advance the purposes of liberal democracy—or whether, on the contrary, they serve more appropriately to support the aims and ambitions of the enemies of democracy.

It is noteworthy that in recent years Lasswell has come to stand for something more than most behavioralists permit themselves. He is no longer content to call his science "value-neutral" and let it go at that. Whatever his earlier ambivalence (between the poles of contemplation and manipulation), he has come to believe that behavioral science is necessarily and staunchly on the side of democracy; almost, it seems, that democracy *is* behavioral science.[117] He has become (in the public image celebrated by David Easton) "Harold Lasswell: Policy Scientist for a Democratic Society."

It may appear that this is less a new position for Lasswell than it is a prodigal's return to an ancestral home. His present perspective would seem, at first glance, to be only that venerable "democratic science" of politics, the peculiar American juxtaposition of natural science with civic virtue, once pre-eminently identified with the name of Lasswell's mentor Charles Merriam.[118] But the situation is not that obvious. Lasswell had long since traveled far beyond the scientific frontiers reached by his predecessors of the progressive era. He had gone on to construct his psychopathology of politics, his descriptive science of power, his political semantics, his skill-politics, and his manipulative theory of elites. After all of this, he cannot so easily go home again. Owing largely to his own efforts, the simple log cabin of early pragmatism is no longer standing; in its place is an Institute for Advanced Research in the Behavioral Sciences, designed in the shape of a pentagon.[119]

Lasswell's many years of assiduous seeking for a reconciliation of science and value—of goal-thinking and trend-thinking, preference and fact, ends and means—would seem to have led by dint of constant incantation to something like the self-fulfilling prophecy. The pursuit of behavioral science has become for him all but synonymous with the pursuit of happiness—and so also with the corollary values of democratic life and liberty. Lasswell is firm in his belief that the union of science and democracy has been consummated; but it is one thing to postulate a relationship, another thing to validate it. The "science of democracy" depends for its initial plausibility upon the prior assumption of what might be called the *democracy of science*: that is to say, the assumption that the methods of natural science carried over rigorously and faithfully to the study of man move unmistakably toward an affirmation of human dignity and personal freedom. But we have already seen that in the hands of its most devoted missionaries, the natural scientists of behavior, this faith in social and political physics has produced with impressive regularity the vision of a techno-scientific future from which all contest and contingency have been removed; and with it a corresponding image of man—manipulated and managed, conditioned and controlled—from whom the intolerable burden of freedom has been lifted.

Bentley: The Politics of Groupthink

> All Nature is but Art, unknown to thee;
> All Chance, Direction, which thou canst not see;
> All Discord, Harmony not understood;
> All partial Evil, universal Good:
> And, spite of Pride, in erring Reason's spite,
> One truth is clear, WHATEVER IS, IS RIGHT.
> —Alexander Pope, *An Essay on Man*

More than half a century has passed since Graham Wallas assisted, as it then seemed, at the birth of a new era in political science with the publication of *Human Nature in Politics*—perhaps the first systematic attempt to apply the findings of modern psychology to the field of government. And it is exactly the same number of years since Arthur F. Bentley produced his own seminal work, *The Process of Government*—perhaps the first systematic attempt (in America at least) to apply the findings of modern sociology to the study of politics. As this suggests, the two books have more in common than the coincidence of their publication in 1908. By general consensus, both have attained the formal rank of "classics" in political science. Both display an aggressively scientific disposition, an almost evangelical zeal to propagate the faith. And both reflect that modern academic temper which has been variously characterized as the pragmatic revolt, the rebellion against formalism, and the Progressive Scholarship.[120]

The most significant common denominator of the Wallas and Bentley studies was their militant protest against the preoccupation of political science with formal institutions and legal-historical analysis—a preoccupation which, as it seemed to both men, either disguised or disregarded the real meaning of politics. By deliberate contrast, theirs was to be a dynamic rather than a static conception of the field. Indifferent to the conventional "morphology" of political structures and institutions, they were excited by the vision of a science of *function*; accordingly they defined the task of the political scientist strictly in empirical terms, as a process of inquiry, in opposition to what they regarded as the arbitrary categories and musty abstractions of the traditional scholarship.

The new political science, as Wallas envisioned it, must penetrate the façade of normative structures and legal frameworks to fix its gaze directly upon the human beings who inhabit them. Likewise the raw material of politics, as Bentley defined it, was to be found in the observable relations and interactions of men—not in the "soul-stuff" and "mind-stuff" of finespun theories thrown up like smoke screens to camouflage the actual facts of social life. Both the Englishman Wallas and the American Bentley were in the vanguard of those restless scholars who sought, at the turn of the century, to fertilize and vivify the ground of political science with the products of the newer social studies—notably psychology and sociology—which were later to form the hard center of the "behavioral sciences." Their two volumes may therefore be regarded as pioneer studies in political behaviorism—classic statements of the prevailing scientific approach to political experience.

Of the two, although Wallas had much the greater immediate impact upon students of politics, it is Bentley whose influence has waxed rather than waned with the passage of time. While few if any political scientists today would admit to taking their departure from the psychological premises of Wallas, a great many claim to do so with regard to Bentley. Moreover, the most striking fact about the present-day adherents of Bentley's "group-basis" theory of politics is how little they have added to the conceptual framework set forth fifty years ago in *The Process of Government*. To review Bentley's contribution is, then, not merely to undertake an exercise in historical exegesis but to come directly to grips with a vigorous and influential movement in contemporary thought.

Bentley has another and peculiar relevance to the present study: he is almost the ideal type of the behaviorist in politics. The coolly objective standpoint from which, in 1908, he looked down upon the Laocoön struggle of the groups was steadily refined and elaborated over four decades into a logical and epistemological system which owed a confessed debt to Watsonian behaviorism as well as to theoretical physics, and which by the mid-thirties its creator was himself describing as "behavioral science."[121] (There is a suggestive thread of progression in the titles of the three books which followed *The Process of Government*: *Relativity in Man and Society, Linguistic Analysis of Mathematics*, and *Behavior, Knowledge, Fact*.)

But if there was change and growth in Bentley's thought over the years, there were also persistent assumptions and recurrent themes. In particular, his commitment to the "observational" perspective of physical science remained undiminished throughout his career, along with what he took to be its indispensable corollary: i.e., the rejection of the individual human being as a meaningful "unit of investigation" for the social sciences, or even as a datum of experience.[122] For Bentley the person was simply an unscientific hypostatization, something just not to be found in actual observation. The scientist could encounter human beings only in active association with one another—that is, only in "groups."[123] It was the conflict and collision of these groups, their reciprocal pressures and mutual repulsions, which was the fundamental reality, the irreducible raw stuff, of social and political analysis.

Some of the currents of thought with which we have already been concerned are to be found curiously intertwined in the behavioral science of Bentley. The ambition of the eighteenth century to reduce all politics to the solid categories of physics; the revitalization of this impulse under the spell of social Darwinism; the rise of the mechanistic psychology of behaviorism, and its subsequent infiltration into sociology and political science —all these familiar movements are rehearsed in Bentley's compact body of published writings.[124] Many of them are sufficiently well-known in their relation to his political theory to need little explanation; but there is one significant source of influence which, although it goes far to illuminate the background and import of Bentley's thought, has received surprisingly little attention.

Although it is customary to classify Bentley among the American rebels against formalism—the academic muckrakers of the Progressive Era—he might with no less accuracy be placed within an older sociological tradition distinctly European in origin and spirit. This was the tradition of *Darwinismus*, which attained its maturity in the group-conflict theories of Gumplowicz and Ratzenhofer. We have seen before (in Chapter 1) how Gumplowicz stood the social Darwinism of Herbert Spencer on its head —reversing its emphasis upon rugged individualism by focusing instead upon the "group" as the chosen agent of cultural evolution. For Ratzenhofer as well, the struggle for survival on the human plane was everywhere and always a struggle between groups: originally those of the tribe, later those which characterize civilized society. Not only was this conflict natural and necessary, it was universal and incessant; Ratzenhofer, indeed, went so far as to formulate a "law of absolute hostility" which held between human groups at all but the very "highest" level of societal development.[125]

In their modern form, as seen by Ratzenhofer and Gumplowicz, the significant groups were simply organized "interests" (shared wants and purposes), each one jealously exclusive in its makeup and actively antagonistic toward all the rest. Yet somehow, this absolute mutual hostility of the group interests (or interest-groups) was seen to lead—whether through conquest, absorption, negotiation, or mere standoff—to an ultimate end-result of adjustment and accommodation. The irrepressible conflict of interests, in short, culminated in an ultimate harmony of interests. Since all this was supposed to occur in strict accordance with natural law, it was obviously not to be tampered with. And since, moreover, the evolutionary process was tacitly identified with the idea of progress, its action must be regarded as not only necessary but beneficent. In effect, whatever emerged from the blind social struggle—whatever natural balance was struck at any given moment—was at once inevitable and progressive. The mechanical equilibrium of the competing interests was synonymous with the public interest.

Paradoxically, then, the "struggle" theory of these social Darwinists was at the same time a theory of "balance."[126] Viewed from one standpoint, the social scene might give the impression of a veritable Gehenna of carnage and conflict—the anarchy of nature red in tooth and claw. Regarded from another standpoint, however, with only a minor shift in perspective, it was transformed into a setting of underlying harmony and serenity: the dynamic equilibrium of universal law and order, guided by the invisible hand of the evolutionary mechanism. For Gumplowicz it was the former viewpoint which predominated; accordingly he looked to the future, rather like Hobbes, with a bleak and foreboding pessimism. In Ratzenhofer's work the focus shifted markedly from the *recurrence* of conflict to the *resolution* of conflict; and so the gloom lifted and the human prospect appeared measurably brighter.

It may have been just this temperamental difference of outlook which persuaded Albion W. Small, an ebullient and pragmatic American sociologist, to choose Ratzenhofer rather than the gloomier Gumplowicz as the model for his *General Sociology*. The central feature of Small's magisterial textbook, which made its appearance in 1905, was a lengthy and loyal adaptation of Ratzenhofer's group theory of society; hence the book served not only to introduce the German sociologist to American students but, by its enthusiastic endorsement, to place his theory of balance-through-struggle squarely in the forefront of American social thought.

It happened that among Professor Small's colleagues at the University of Chicago at the turn of the century was the youthful

Arthur Bentley, then serving an apprenticeship as a docent in sociology.[127] Almost a decade before, Bentley had gained a first-hand acquaintance with Ratzenhofer and Gumplowicz in the course of graduate study at Heidelberg, where he also attended the seminars of another group-conflict theorist, Georg Simmel. The personal influence of Small (who was an unusually gifted teacher), as well as of the Europeans whom Small so ardently championed, is evident throughout *The Process of Government*.[128] Much that otherwise would remain ambiguous or enigmatic in Bentley's political science becomes clearer in the light of this early training and experience; and indeed much of what has been considered most original and creative in his work emerges—when properly situated in the line of scholarly succession—as derivative and even conventional.

Thus the process of politics, for Bentley, was the struggle of the groups for the stakes of power, neither more nor less: a struggle which, however carnal or cannibalistic it might appear, was to be examined with a zoologist's indifference to the cogency of competing claims. If the combatants were of unequal strength, at least they were of equal standing in the eyes of the scientific observer. Indeed his neutrality was reinforced by another premise: there were no meanings, nothing of significance, to be reckoned with either *behind* the interests (at the personal level) or *beyond* them (at the public level). Like the German Darwinists, Bentley made an elaborate point of throwing out the individual and his cherished values as something irrelevant and incompetent. In terms remarkably parallel to those of Watson, he repudiated such "spooks" as consciousness, reason, purpose, ideas and feelings as the creatures of a metaphysical nightmare from which we should be trying to awake.[129] Nor on the other hand could there be any appeal beyond the colliding interests to "an interest of the society as a whole"; for that too was a spook, which careful analysis would reveal as mere camouflage for still further parochial and self-serving interests.[130]

With the exponents of *Darwinismus*, also, Bentley contrived to get rid of individualism while retaining the atomistic world-view which had undergirded it. That his political science owed as much to Newton as to Darwin is evident from Bentley's insistent espousal of what he took to be "the naive viewpoint of physics," along with its freight of mechanistic metaphor. Like Marx, he envisioned nothing less than the systematic application to society of the laws of matter and motion.[131] Like Saint-Simon, he believed it all to be explainable in terms of universal gravity— of attraction and repulsion, force and counterforce, pressure and resistance. And like the economic mechanists who took their cue from Townsend, he perceived the political market as automatically self-regulating—forever resolving itself into a steady state of dynamic equilibrium.[132]

The implications of so thoroughgoing a scientism may be witnessed in the inflection given by Bentley and others to the concept of *force* or *pressure*. As the mechanist uses it this key notion replaces and drives out the alternative idea of *purpose*, in the ordinary sense of conscious effort and intention. It is force and force alone, applied from without, which provides the means of explaining causal relationships and behavioral effects. As one critic of the scheme has written, "Since this theoretical model is that of a machine which responds or adjusts to external force, there would be no initial motion in the system without the application of external force."[133] Just so Bentley defined his groups and their activity strictly in terms of reciprocal pressures, of force calling out force—as in Ratzenhofer's law of "absolute hostility."[134] Not only are inner purposes meaningless in such a system; the very "reasons" that men persist in uttering are only rationalizations after the fact of external force—and its equally mechanical counterforce.

It is conceivable that Newton himself might have been embarrassed by this straightforward extrapolation from his Third Law; for, as Bronowski has remarked, in his conception of gravitation as a "force" (no longer a tenable notion in physics) Newton was actually imputing to inorganic nature the very quality of human effort or will.[135] What Bentley and his epigones have done is to reverse the analogy: their hypothesis imputes to human activity the quality of gravitational force. But the "force" they have taken over from nature is only the "force" which man himself had put into it.[136] With, however, one difference: in the course of its hegira the concept of force has undergone a sea-change. Having started out in recognizably human form, as volitional and purposive, it has returned dehumanized and empty—as mere "blind force."

The blindness of the forces at work in this political cosmos is matched by the self-inflicted blindness, or near-blindness, of the scientific observer. He is permitted to perceive only what lies directly in his line of sight: the ongoing activity in the foreground, the immediate "parallelogram of forces." For him there is no past, except as it is caught up and made manifest in the present; there is no future, except as it is immanent in the here-and-now. In effect the political process has come out of nowhere and is going nowhere; history is bunk, and Whirl is king. This is the constricted angle of vision which would seem almost willfully to be enforced by the "observational" perspective, the naive viewpoint of Bentley's physics. The camera eye of the scientist, at the moment of observation, artificially stops the action and fixes the wheeling groups in their orbits while the image is recorded. When the film has been developed—when, as Bentley put it, "the description is complete"—it turns out, inevitably, to be in perfect balance and focus. To see it otherwise, to detect any disproportion or imbalance in the recorded scene, would be to smuggle in a valuation which itself reflects an "interest." But the scholar must be, to the best of his ability, disinterested—which is to say that he may see no evil, hear no evil, and speak no evil.

It does not appear to be recognized by those students who have embraced Bentley's theory of politics that the presumption of "equilibrium," imposed by their value-neutralism as well as by their ingenuous physics, itself conveys a definite judgment—in favor of the interests which hold the field. That this "cynical conservatism"[137] is unintended does not affect the consequence: which is that right and justice, by the natural law of science, are on the side of the bigger battalions. The very finding of a "balance" is, of course, an interpretation which distinctly plays favorites. Since at any moment some are winning and some are losing, one man's (or one group's) equilibrium is another's *dis*equilibrium.

In seeking to assess the error in all this it is instructive to recall that the "group-basis" theory of politics, as it has come to be designated, did not originate with Bentley or with his predecessors of German Darwinism.[138] What they contributed, for all its novelty, was essentially a variation on an old theme: that of political pluralism. But the scientific twist they gave to it has made that theme all but unrecognizable. The traditional theory of the contest and interplay of group interests was not so much a scientific description as a normative injunction; in fact it contained the moral basis of liberal democracy. In order for the truth to be known, this theory argued, speech must be free; in order for wise decisions to be made, all the interests must be in the field and all the values articulated. The ancestry of this liberal pluralism might well be traced, beyond Mill and Madison and Montesquieu, all the way to the dialectical principle of Socrates: the method of verbal contest or discussion grounded on the faith that there was indeed a truth to be reached, through mutual deliberation, on which reasonable men could agree—or, expressed another way, that there existed a common "interest of the society as a whole" which it was the object of the dialogue to ascertain. It was never supposed that this dialectical process—the conversation of viewpoints—was something which sprang up spontaneously, proceeded mechanically and must always end happily; nor was there any doubt that its uncertain career might at any moment be deflected and derailed, through the sophistry of rhetoric or the simple failure of nerve. In short, there was no invisible hand in the background to assure that all Discord was really Harmony not understood, and all partial Evil universal Good. Then as now, it was not the tolerant Socratics but the impatient disciples of power—men like Callicles and Thrasymachus—who committed the naturalistic-scientistic fallacy, and whose arguments were reducible to the conviction that justice is the interest of the stronger.[139]

It may seem at first glance that this group-behavioral approach, with its Panglossian acceptance of political things as they are, bears little resemblance to the manipulative urges of those other prominent forms of social behaviorism whose visions of a techno-scientific utopia we have reviewed earlier. Indeed there are differences; but there are also points of convergence. If it is true that an unanticipated consequence of Bentley's analysis has been to bestow "academic absolution" (in the trenchant phrase of Peter H. Odegard) upon the group struggle and its bloodied survivors, an equally patent result has been to focus the attention of both participants and observers upon the strategies and logistics of how to succeed in politics—without really caring.[140] For when "the description is complete," prediction is in order; and with the possibility of prediction, as any scientist knows, goes the power and opportunity of control. It is exactly this combination of indifferent empiricism with practical manipulation which, as we have repeatedly observed, has characterized the enterprise of behavioral science in its most consistent and conventional examples.

There is a very suggestive analogy to be found in the persistent habit among tough-minded behavioralists of referring to politics as a "game" and to its underlying commitments as the "rules of the game."[141] For the concept of the *game*, in its traditional and intrinsic meaning, is that of a mock contest, a diversionary pastime definitely set apart from the serious and responsible affairs of real life. By its playful counterfeiting of reality, the game shows itself to be the essence of unreality—that is, fantasy. In light of this it is difficult not to infer that the "game" of politics, as it appears through the looking-glass of the convinced behavioralist, is just such a wonderland of illusion and make-believe—in which cardboard heads may roll and walruses may weep but a rational child will remain undeceived. For it is really a trivial roundelay; its very point is to be pointless. The game is its own reward, to be played (and kibitzed) for its own sake without the irrelevant intrusion of higher purposes or continuing consequences; the rules of the game are arbitrary and capricious, and if they must always be formally enforced they may also be artfully circumvented.

No doubt this stretching of the game analogy must seem unfair when applied in criticism of responsible scholars whose work is uniformly marked by seriousness of purpose as well as by sophistication of design.[142] But observe the statement of a British representative of the group-process school:

> I wonder if I can make this conception clearer by saying how I have always privately pictured the political process? You all know those large, electric pin-tables in amusement saloons? The pins carry different scores and every time your ball hits a pin, a score flashes up on an electric indicator. In my analogy the ball is the social issue to be decided. It is shot into the field of play and at once is tossed from one pin to another. It bounds and rebounds, running the gauntlet of organized attractions and repulsions. The score, i.e., the final outcome, depends on the tilt of the table. And it also depends on the pins it falls against. As it proceeds down the field of play, so the score flashes up, bit by bit, until in the end the ball runs out and the score is complete. . . . So it seems to me in the social field.[143]

In this graphic image of the political process, the analogy of the playground (or the amusement saloon) betrays a deeper predilection. The game of political pinball is frankly and thoroughly mechanical: it is the modern electronic equivalent of that billiard-ball universe of blindly bumping atoms familiar to the last century; and the only significantly operative rules of the game are the cosmic laws of motion and inertia. It is easy to see the drastic effect which this truly "naive viewpoint of physics" must have upon the ancestral faith of democracy that the governmental process is affected with a public interest which is not just an "absent" or "unorganized" group pressure but the interest of the community as a whole—organized, present and accounted for in the primary association of the body politic. "Group theorists, like the early radical behaviorists," writes Odegard, "have all but banished reason, knowledge, and intelligence from the governmental process."[144] It may be added that in doing so they have destroyed any basis on which one process or system of government might be judged preferable to another, except that of the struggle for survival—and any basis on which involvement or concern with the activity of politics might be vindicated, except that of immediate advantage. In reducing the individual to the group, they have lost the perception of those democratic values (among them dignity and freedom) which spring from a passionate regard for the human person. And in reducing the public interest to the level of the private pressure group, they have lost the thread of that essential tradition of Western thought which has given us a sense of the public philosophy and an intimation of the good society. As the distinguished scientist-philosopher Michael Polanyi has warned us,

> Man is strong, noble and wonderful so long as he fears the voices of this firmament; but he dissolves their power over himself and his own powers gained through obeying them, if he turns back and examines what he respects in a detached manner. Then law is no more than what the courts will decide, art but an emollient of nerves, morality but a convention, tradition but an inertia, God but a psychological necessity. Then man dominates a world in which he himself does not exist. For with his obligations he has lost his voice and his hope, and been left behind meaningless to himself.[145]

Humanization—from Physics to Politics

CHAPTER IV

An Uncertain Trumpet:
The New Physics

Unfortunately the supplements to science which most philosophers supply in our day are not conceived in a scientific spirit. Instead of anticipating the physics of the future they cling to the physics of the past. They do not stimulate us by a picture, however fanciful, of what the analogies of nature and politics actually point to; they seek rather to patch and dislocate current physics with some ancient myth, once the best physics obtainable, from which they have not learned to extricate their affections.
—George Santayana, *The Life of Reason* (1906)[1]

DURING THE THREE HUNDRED YEARS in which the world view of Newtonian physics dominated the stage of Western thought, there were always voices of dissent. The most prominent and authoritative, through much of the period, was of course that of the church: the organized voice of the spirit, which had recognized its own jeopardy almost from the first appearance of the "infernal machinists."[2] There were, besides this, recurrent movements of resistance led by humanists who feared with John Donne the loss of all coherence and "just supply" from the scientific disintegration of the medieval synthesis. In various fields of thought the reign of mechanism was successively challenged by competing visions—such as those of vitalism, romanticism and idealism—all more or less reflecting, despite their other differences, what William James was to call the "tender-minded" (as opposed to the "tough-minded") attitude toward the meaning of experience and reality.[3]

After the seventeenth century, however, these humanist alarms and excursions proved more and more ineffectual in slowing the march of the mechanistic philosophy toward the *hubris* of an all-embracing and all-sufficient scientism. The protestants were, after all, unscientific outsiders, often the apologists for vested interests and commonly the defenders of a rear-guard position which appeared fatally compromised by the burden of an untenable past. To oppose the prevailing cosmology was simply to be against "science"; and to be against science was to stand squarely in the path of truth and progress. Even those few intrepid spirits —among them such venerated figures as Goethe and Hegel—who ventured to challenge the orthodoxy from an alternative scientific standpoint were roughly repudiated and driven from the field.[4]

But in the twentieth century, for the first time, the image of the great machine came under direct attack by science itself— and not merely by a scattering of isolated researchers but by the full mainstream of theoretical and experimental physics. The profound significance of this assault from within upon the framework of the traditional world view has been well underscored by one of its major protagonists, Werner Heisenberg. Pointing out that many previous attempts had been made to "get away from this rigid frame" which seemed too narrow to embrace the heart of physical reality, he observed:

> But it had not been possible to see what could be wrong with the fundamental concepts like matter, space, time and causality that had been so extremely successful in the history of science. Only experimental research itself, carried out with all the refined equipment that technical science could offer, and its mathematical interpretation, provided the basis for a critical analysis—or, one may say, enforced the critical analysis—of these concepts, and finally resulted in the dissolution of the rigid frame.[5]

The Decline of Mechanism

What has since come to be known as the revolution in modern science reached its climax in the nineteen-twenties and thirties in the field of quantum physics; but its preliminary engagements had been fought and won half a century earlier. Even before that, the Laplacean ideal of absolute certainty had been disturbed by the development of thermodynamics, with its reliance upon probability.[6] But the first substantial indication of a crack in the mechanical model appeared in the final quarter of the century in the wake of those exhaustive investigations into electricity and magnetism through which, for upwards of a hundred years, prodigies of scientific ingenuity had been expended in a futile effort to fit these refractory phenomena into the framework of classical mechanics. "The great change was brought about," as Einstein has noted, "by Faraday, Maxwell and Hertz—as a matter of fact half unconsciously and against their will."[7] In 1873, proceeding from the ordained mechanical premises, Clerk Maxwell ended up with a set of equations (the electromagnetic theory) which clearly repudiated those premises and cast doubt upon the entire foundation of Newtonian mechanics. Even then, many leading physicists rejected Maxwell's conclusions because they could not be rendered visible and substantial in terms of a mechanical model.[8] It was only after decades had passed that the radical implications of Maxwell's theory were fully assimilated by his scientific colleagues, for whom the ideal of mechanism had been too long an unquestioned article of faith.[9] In defense of his thesis, Maxwell himself was brought to an extraordinary intuition of things to come:

If . . . cultivators of science . . . are led to the study of the singularities and instabilities, rather than the continuities of things, the promotion of natural knowledge may tend to remove that prejudice in favor of determinism which seems to arise from assuming that the physical science of the future is a mere magnified image of that of the past.[10]

The success of the electromagnetic theory was significant not only as the first breach in the great wall of mechanism, but also as a premonitory clue to what might be termed the "postmodern" conception of the nature of scientific knowledge. Previously it had been assumed by the faithful that the descriptions of which science was capable encompassed the whole of knowable reality—in other words, that what escaped the net of the scientific mechanist either did not exist or was not worth knowing.[11] Indeed, scientific explanation not merely ordered and described the objects of its concern but penetrated to their very essence. "Nothing was uncertain for it": nothing (to paraphrase a line of Santayana) could withstand the centrifugal ray of its intelligence, darting from the slime to the stars. Through its methods of mechanical reduction and mathematical reproduction, as well as by its postulate of neutral objectivity, modern science sought for (and was on the way to finding) not a limited aspect of reality but the whole truth and nothing but the truth.[12] It was this passionate quest for certainty, this unqualified faith in the power of "Natural Magick" to expose and explain the deepest secrets of the universe, which principally characterized the creed of scientism and constituted its peculiar *hubris*.[13]

And it was this fundamentalist faith which was profoundly shaken, if not yet overturned, by the failure of the mechanistic assumptions to account for the nature of electricity, and by the incontrovertible success of Maxwell's effort to comprehend the mystery in terms basically alien to the Newtonian perspective. As a result of Maxwell's equations, as the mathematician Sullivan has put it, "All that we knew about electricity was the way it affected our measuring instruments. . . . The precise description of this behaviour gave us the mathematical specification of electricity and this, in truth, was all we knew about it."[14] The significance of this departure from the standard procedure and expectation of classical mechanics dawned slowly upon the world of science. Natural scientists were understandably reluctant to modify a framework which had proved so continuously fruitful both in theory and practice; and not only natural scientists but social and

political scientists as well were unwilling—and in substantial numbers remain unwilling today—to abandon a world view which, at
whatever the price to human dignity and moral freedom, had
brought determinate order and the illusion of certainty to the
once-mysterious universe.[15] Writing more than half a century after
Maxwell's discovery, Sullivan drew the moral of his achievement:

> It is only now, in retrospect, that we can see how signifi-
> cant a step this was. . . . *It has become evident that, so far as the
> science of physics is concerned, we do not require to know the
> entities we discuss, but only their mathematical structure.* And, in
> truth, that is all we do know. It is now realized that this is all the
> scientific knowledge we have even of the familiar Newtonian
> entities. Our persuasion that we knew them in some exception-
> ally intimate manner was an illusion.[16]

If the full significance of Maxwell's equations remained
unappreciated by the generality of physicists in his time, there
were nevertheless a few who read the signs correctly and who
ventured to resist the dogma of omniscient scientism. Appropriately enough, it was Heinrich Hertz—the scientist who more
than any other had confirmed and rounded out the work of Maxwell—who most closely anticipated what Heisenberg was to call
the "essential insight" of twentieth-century physics. In his *Principles of Mechanics* (1894), Hertz argued that the propositions of
natural science are only as valid as the limited aspects of nature
which they purport to describe, and that science is not a philosophy projecting an explanation of nature as a whole or of the
essence of things.[17] A decade earlier Ernst Mach had made substantially the same point in his *Science of Mechanics*;[18] and no commentator before or since has put the case for the "limitations of
science" with greater clarity or pungency than Karl Pearson,
whose classic *Grammar of Science* made its first appearance in
1892.[19]

Little of this attack upon the pretensions of scientism, to be
sure, was recognized by the body of natural scientists at the turn of
the century. But what had come to be accepted was already considerable: i.e., that the Newtonian principle of actions at a distance,

which lay at the heart of the mechanistic world view, was inadequate to account for the interactions between fields of force demonstrated by the equations of Maxwell, in which the electromagnetic field could exist as a wave independent of the material source. The implications of this breakthrough were soon evident. "What was true for electrical action," as Einstein has recalled, "could not be denied for gravitation. Everywhere Newton's actions-at-a-distance gave way to fields spreading with finite velocity."[20] Everywhere, in short, what had hitherto been regarded as an indispensable cornerstone of the traditional cosmology was quietly and permanently rolled aside.

The decline of mechanism which these developments carried in their train is nowhere more graphically illustrated than in the fate of the ether (or Aether): that logical "end product of Newtonian physics"[21] which at once epitomizes the world view of classical mechanics and underlines its failure. Introduced at least as early as the seventeenth century, the ether was the *deus ex machina* designed to set at rest all the unsolved riddles of the Newtonian universe. Among other things it was the "invisible medium in which the stars wandered and through which light traveled like vibrations in a bowl of jelly."[22] It furnished a mechanical model for every occurrence in nature, and it provided the ultimate frame of reference, the "absolute and immovable space," which Newton's cosmology required. This "monster," as Sullivan has labeled the ether construct ("the most unsatisfactory and wasteful product of human ingenuity that science has to show")[23] became through an endless series of incarnations the fundamental explanatory principle, or court of last resort, to which mechanists might appeal whenever their experimental models broke down or their measurements failed to tally. In the nineteenth century, as Jeans has put it, the "ether habit" took hold of scientists until there were nearly as many ethers as there were unsolved problems in physics.[24]

So long as the rigid framework of classical physics remained unchallenged, it could scarcely be otherwise; wherever a mechanical explanation could not be found through observation, it was only necessary to invent it. The ingenuity of the inventors, extended over several generations, seemed boundless. "Ethers were invented," remarked Maxwell at the height of the fashion, "for the planets to swim in, to constitute electric atmosphere and magnetic effluvia, to convey sensations from one part of our body to another, till all space was filled several times over with ether."[25] For a time, following Maxwell's own discovery, it was still possible for eminent scientists such as Lord Kelvin to maintain that nothing was so certain in physics as the reality of the luminiferous ether.[26] But by the turn of the century—even before the special theory of relativity arrived to banish the monster for all time—most physicists had grown weary of the "ether habit" and disillusioned with its possibilities. "The construction of ethers became a decaying industry, and largely because there was so little demand for the product."[27] With regard to Maxwell's theory in particular, physicists came gradually to accept the new field concepts as irreducible and, in the words of Einstein, resigned themselves to giving up the idea of a mechanical (or ethereal) foundation.

But apart from their potential implications for epistemology, none of these new departures of the nineteenth century—thermodynamics, the electromagnetic theory, the rejection of the ether—could properly be described as revolutionary. Even though in the explanation of electricity "matter" had been replaced by "fields of force," the interactions between fields could still be viewed as taking place in Newtonian space and time, in accordance with accepted laws. Moreover, they could still be described with the requisite objectivity (i.e., without regard for the manner of their observation); and, finally, they were still dependent upon the ancient foundation of atoms, which remained as ever the indivisible building stones of the material universe.[28]

The publication of Einstein's special theory of relativity in 1905, followed a decade later by his general theory, signified a radical extension of these departures from the mechanical framework of classical physics. With the appearance of relativity, "the study of the inner workings of nature passed from the engineer-scientist to the mathematician."[29] In the new theories gravitation was no longer regarded as a mechanical "force," but instead took on the character of a mathematical formula governing the curvature of space and the acceleration of moving bodies. Space and time—those formidable absolutes of the common-sense world—lost both their absoluteness and their independence, as they were welded into a single four-dimensional continuum of space-time. No doubt there was still a crucial sense in which these modifications of the old world view, for all their profound and far-reaching effects, must yet be regarded as brilliant corrections rather than as fundamental contradictions of the main tenets of classical physics.[30] But if the mechanistic universe was not yet overthrown, it was surely altered (not to say disfigured) beyond easy recognition. Not "matter" but "energy" was now the basic datum of science; no reliance could henceforth be placed upon actions-at-a-distance, nor upon mechanical conceptions of force or of quasi-solid ethers, nor upon the integrity and stability of Space and Time as familiarly conceived—nor, indeed, upon the bedrock axioms of Euclidean geometry.[31] "The progress of science," declared Whitehead in 1925,

> has now reached a turning point. The stable foundations of physics have broken up. . . . The old foundations of scientific thought are becoming unintelligible. Time, space, matter, material, ether, electricity, mechanism, organism, configuration, structure, pattern, function, all require reinterpretation. What is the sense of talking about a mechanical explanation when you do not know what you mean by mechanics?[32]

It was not only that physicists no longer knew what they meant by mechanics; in more and more corners of their field, they were ceasing to think at all in mechanical terms. As they had outgrown the "ether habit," so they were discarding the underlying framework of mechanistic explanation—that common-sense perspective born perhaps of the primitive animism which invested nature with demonic forces and lively pressures, in combination with the modern tendency to extrapolate from the everyday world of technics to the data of science. "Our modern minds have, I think," remarked Jeans, "a bias towards mechanical interpretations. Part may be due to our early scientific training; part perhaps to our continually seeing everyday objects behaving in a mechanical way, so that a mechanical explanation looks natural and is easily comprehended."[33] Moreover, in the pre-relativistic era, with allowance for a few lacunae, the mechanical explanations clearly "worked"; that is, they gave an account of natural occurrences adequate to satisfy the questions which were being asked of nature. (It was of little consequence to scientists then that the answers were, as later came to be recognized, only rough approximations; they fulfilled their pragmatic purpose, and in ordinary cases continue to do so.)[34] From Galileo to Kelvin men of science "looked abroad on the universe and saw lumps of matter moving about in accordance with certain laws. Both the lumps of matter and the laws were supposed to exist quite independently of our minds. We simply saw what happened to be there."[35]

After the full impact of relativity had been felt, however, little was left of this ingenuous belief in the scientific image of objective reality, as faithfully registered by the senses. It began to be recognized that the familiar picture, far from being a genuine photographic reproduction of an independent reality "out there," was rather more on the order of a painting: a subjective creation of the mind, which could convey a "likeness" but could never produce a replica. The old distinction between primary and secondary qualities began to lose all clarity and conviction, for very much the reasons that Robert Boyle had advanced nearly three centuries before.[36] The formulations of Einstein made clear that "even space and time are forms of intuition, which can no more be divorced from consciousness than can our concepts of color, shape, or size. Space has no objective reality except as an order or arrangement of the objects we perceive in it, and time has no independent existence apart from the order of events by which we measure it."[37] Before the avalanche of fresh speculation set in motion by relativity had moved very far, Sir Arthur Eddington—proceeding with rigorous mathematical logic from the premises of Einstein—arrived at the unsettling conclusion that "all the laws of nature that are usually classed as fundamental can be foreseen wholly from epistemological considerations. They correspond to *a priori* knowledge, and are therefore *wholly subjective*."[38] And Sir James Jeans, scarcely less renowned for his solid contributions to physical knowledge, felt safe in declaring:

> Today there is a wide measure of agreement which on the physical side of science approaches almost to unanimity, that the stream of knowledge is heading towards a non-mechanical reality; the universe begins to look more like a great thought than like a great machine. Mind no longer appears as an accidental intruder into the realm of matter; we are beginning to suspect that we ought rather to hail it as the creator and governor of the realm of matter. . . .[39]

The Quantum Revolution

The entire series of advances and departures in physical theory thus far described, for all their radical originality and influence, may be fairly regarded as only a prologue to the real revolution in twentieth-century science: that of quantum physics.[40] After all else has been duly recognized and credited, it was finally in the submicroscopic interstices of the atom—rather than in the telescopic reaches of outer space or the macroscopic world of everyday experience—that the primary postulates and "necessary truths" of the Newtonian cosmology were one by one brought under critical scrutiny, and one by one found wanting.

Although the quantum theory was originally formulated by Max Planck at the turn of the century, its startling hypothesis— that energy is emitted not in the continuous stream of common supposition but in discontinuous packets or *quanta*[41]—was only gradually and grudgingly acknowledged by his fellow physicists. The reason was not simply, as in the case of Maxwell, that the quantum of energy resisted representation in terms of a mechanical model; more disturbingly, its assertion of discontinuity at the bottom of things rudely violated the cardinal faith of modern science in the uniformity and continuity of nature—specifically, its belief that the evolution of every self-contained physical system is constituted by a continuous chain of causally related events.[42] Only after "Planck's constant" and its theoretical underpinning had proved successful in a number of physical applications—most notably in Einstein's extension of the principle to all forms of radiant energy through the establishment of the photoelectric effect[43]—was the new conception fully accepted as a fundamental law of physical science.

The extent of the revolution which this discovery ushered in is suggested by the observation of de Broglie that, "on the day when quanta, surreptitiously, were introduced tho vast and grandiose edifice of classical physics found itself shaken to its very foundations. In the history of the intellectual world there have been few upheavals comparable to this."[44] Through the next half-century the quantum theory steadily made its way into nearly every department of physical science. In addition to Einstein's Nobel Prize-winning researches, the principle found a variety of fruitful applications in chemistry; it shed light upon the Third Law of Thermodynamics; it modified the kinetic theory of gases; it was (most significantly of all) applied by Bohr to explain the structure of atoms, and later by Dirac in his prediction of the positron. Even Maxwell's equations underwent reappraisal in an effort to adjust them to the new ideas. More recently, there have been public suggestions that "space" itself may be found to adhere to the quantum principle—a conception whose reverberations may conceivably outdistance all previous extensions of the theory.[45] "In short," as one historian summarizes, "optics, mechanics, thermodynamics, chemistry, the statistical laws, and many others have one by one come under the sway of the quantum theory. Needless to say, our entire outlook on the physical world has been affected."[46]

The hypothesis of energy radiation as finite and discontinuous, which was central to the quantum theory from the beginning, shook the confidence of scientists in the order and continuity of natural processes. But it was not until 1913 that the first major offensive of quantum physics against the traditional cosmology—namely, its attack upon the deterministic conception of physical causality—was launched with the publication of Niels Bohr's theory of the atom, which demonstrated that the entire internal organization of matter rests upon the existence of quanta.[47] Bohr's atom represented the most complete break with classical physics yet accomplished, and, through its hypothesis of the unpredictability of individual atomic events, provided "the first set of arguments in favour of indeterminism" in the fundamental operations of nature.[48] Later experiments led to the displacement of Bohr's pioneer theory, but the rupture with traditional mechanics which it had signalled, far from being repaired, was progressively widened and reinforced in the subsequent development of quantum physics —not least of all through the creative contributions of Bohr himself.[49]

Uncertainty: The Eclipse of Mechanism

In the course of this development two crucial principles came to be established which together serve to illuminate the total eclipse of mechanistic determinism and objectivism enforced by quantum physics: namely, Heisenberg's principle of uncertainty (or indeterminacy), and Bohr's principle of complementarity. The first of these, in particular, stands in stark opposition to the basic causal thesis of classical physics that the evolution of every isolated mechanical system is rigorously determined by its initial state; and, more specifically, that exact prediction of its future behavior follows directly upon knowledge of its present position and velocity.[50] Originating with Newton's celestial mechanics, this rigorously deterministic doctrine had soon been extended to all fields of physical science (and, by eager extrapolation, to the social sciences as well). "More or less consciously," as de Broglie has said, "the inner determinism of natural phenomena, implying their complete predictability at least in principle, had become a kind of scientific dogma."[51] To be sure, it was conceded that our knowledge of the twin conditions of predictability is often incomplete; but that did not affect the principle involved, and indeed it was believed that in the not distant future complete information and absolute precision would

in fact be achieved. Again, it was recognized by classical physicists
that no mechanical system (except possibly the universe itself) is
perfectly isolated; but here as well the principle of absolute objective detachment and determination remained a valid goal. The
supposition, in the words of Max Born, was "that the external
world, the object of natural science, and we, the observing,
measuring, calculating subjects, are perfectly separated, that there
is a way of obtaining information without interfering with the
phenomena."[52]

The principle of uncertainty, first mathematically expressed
by Heisenberg in 1927, declared on the contrary that precisely the
kind of information which was assumed by classical physics as the
prerequisite to exact prediction—i.e., the simultaneous knowledge
of position and velocity—was impossible of attainment in microphysics. The relationship of the two variables is such that the more
accurately we measure the one the less accurately are we able to
define the other; and this degree of uncertainty is governed by an
irreducible minimum.[53] Since the state of a particle is defined by its
position and velocity together, we can never know exactly what
that state is; consequently, we can never decide whether a given
initial state determines subsequent states, and therefore the existence of rigorous causal connections and laws cannot be tested at all.
In short, we can never be certain of the future because we are never
in fact quite sure of the present.[54]

In the new dispensation, of course, we are as well equipped
as ever to answer particular experimental questions taken one by
one, such as the position (*or* velocity *or* wave length) of a particle.
What we can no longer hope to do is to "weave together these isolated segments into the web of perceptible and causal relations
which alone constitutes the permanent objective nature of the
classical scientific world view with its unchangeable 'things.' "[55]
We may choose which segment of the whole we wish to comprehend; but the whole itself eludes the grasp of the measurer. This is,
as Oppenheimer has emphasized, a very different view of reality
from that of Newton's great machine: "It is not causal; there is no
complete causal determination of the future on the basis of available

knowledge of the present. It means that every intervention to make a measurement, to study what is going on in the atomic world, creates, despite all the universal order of this world, a new, a unique, not fully predictable, situation."[56]

This is not to say, of course, that the uncertainty relations of Heisenberg forbid the possibility of all prediction in the atomic realm; but only that the foresight which is open to us is solely that of probability rather than of certainty—a difference in degree which is ultimately a difference in kind. The old deterministic laws of the giant machine have given way, not indeed to "lawlessness" but to statistical calculations—to the laws of chance. Today if we wish to read the fortune of a particle we must, as Bronowski observes, "allow it to have some uncertainty: some range of alternatives, some slack—what engineers call some tolerance."[57] And, of course, "once we have any uncertainty in prediction, in however small and distant a corner of the world, then the future is essentially uncertain—although it may remain overwhelmingly probable."[58]

It is precisely this note of ultimate contingency, shown by Heisenberg to lie inescapably at the heart of matter, that has induced most theoretical and experimental physicists over the past thirty years to abandon a conviction which the cumulative scientific experience of three centuries had erected into the dimensions of incontestable dogma. Not all the scientists, to be sure, were wholly pleased with this abrupt displacement of the ancestral faith which quantum physics carried in its train. While the majority were, like Max Born, heartened by the demise of a mechanical model which had seemed to convert all existence into automation,[59] there were others (ironically including Einstein and Planck) who, although unable to refute the claim of the new theories, frankly looked forward to the eventual restoration of determinate law and order in the universe.[60] "I do not think," concluded Eddington in the early thirties, "that there is any serious division of opinion as to the decease of determinism. If there is a division among scientists it is between the mourners and the jubilants."[61]

Possibly even more significant in its epistemological implications than this monumental breakthrough of indeterminacy was the reason which Heisenberg's principle adduced to account for it:

namely, *"the fact that we cannot observe the course of nature without disturbing it."*[62] It turned out, to put it simply, that the measuring instruments of microphysics themselves exert a distinct influence upon that which they seek to measure, thus rendering its behavior in one or another respect unpredictable. The very attempt to observe a particle "knocks it off its course," and the more accurately we pin down its position, for example, the more unsure we are of the degree to which we have affected its momentum. Among atomic physicists, in short, there are no innocent bystanders; the act of observation is at the same time unavoidably an act of participation. As Andrade has succinctly put it: "Observation means interference with what we are observing. . . . Observation disturbs reality."[63]

At the base of the classic Newtonian world view, as we have seen in Chapter 1, was the conception of Nature as an independent reality objectively apart from man and therefore objectively observable without reference either to the means of observation or the singularity of the observer.[64] The image of Fontenelle, the vision of Laplace—and prior to them both the dream of Descartes—depended for their initial plausibility upon this firm dichotomy of man and nature, of subject and object, through which alone the exact description of the great machine became a possibility. "The objects with which the world was filled," in the words of Oppenheimer, ". . . were found by observation and by experiment; but it would have occurred to no one that their existence and their properties could be qualified or affected by the observations that told of them."[65] Man—and pre-eminently scientific man—was only a mechanically-minded spectator at the grand performance of nature, unaware that the monkey-wrench he carried and was bound to throw would have its distinctive effect upon the proceedings.

The principal lesson derived by quantum physicists from the discoveries of the past half-century is one which is addressed directly to this venerable ideal. In the famous figure of Bohr, it is simply that man is at once an actor and a spectator in the drama of existence; or, in the more elaborate analogy of Max Born,

We may compare the observer of a physical phenomenon not with the audience of a theatrical performance, but with that of a football game where the act of watching, accompanied by applauding or hissing, has a marked influence on the speed and concentration of the players, and thus on what is watched.[66]

Expressed more directly, the principle of uncertainty has closed the artificial gap between the knower and the known which had been held open by the combined exertions of philosophers and scientists since Galileo, but was already narrowed appreciably by the insight of relativity theory that the thing observed is intimately bound up with the observer and his observation.[67] Nor can it be maintained that only the disembodied measuring instruments, and not the observing scientist, occasion the interference and bring about the disturbance; for the decision of the scientist himself, the subjective choice he has to make, is inextricably involved in the outcome.[68] And in the last analysis the interpretation given to it all, the very meaning of the observation, is peculiarly his responsibility. "We wanted to press on behind appearances to the things themselves," Weizsäcker has written, "in order to know them and to possess them; now it appears that precisely beyond our natural perceptual world the very concept of *thing* can be defined only in relation to the man to whom it appears or who himself makes it. . . . Contemporary physics compels the physicist to look upon himself as a subject."[69]

What this means is that the focus of scientific explanation has shifted from the "thing itself," from Nature as an independent reality "out there," to man's own observation of nature. The author of the uncertainty principle has lucidly described the new perspective:

. . . we can no longer consider "in themselves" those building-stones of matter which we originally held to be the last objective reality. This is so because they defy all forms of objective location in space and time, and since basically it is always our *knowledge* of these particles alone which we can make the object of science. . . . From the very start we are involved in the argument between nature and man in which science plays only a part, so that the common division of the world into subject and object, inner world and outer world, body and soul, is no longer adequate and leads us into difficulties. Thus even in science the object of research is no longer nature itself, but man's investigation of nature. Here, again, man confronts himself alone.[70]

This modern (or postmodern) image of the scientist as *actor*, as "participant-observer" rather than detached spectator, has led to renewed consideration of the process of observation as a form of interaction or "transaction"—terms which stress the fact that contributions are being made from both sides of the measuring apparatus. Thus Bridgman has asserted that "the mere act of *giving meaning* through observation to any physical property of a thing involves a certain minimum amount of interaction"—and indeed that this implies not merely a definite action on the part of the observer but one that entails "certain universal consequences."[71] The alternative term "transaction," proposed by Dewey and Bentley, conveys still more sharply this reciprocity of influence on the part of the knower and the known in the process of scientific inquiry.[72] We shall have occasion later on to consider the significance of the shift in scientific thought away from classical detachment toward heightened self-awareness—more specifically, toward the recognition by the observer of his own ubiquitous and ineluctable engagement. For the present it may be sufficient to underscore its essential point through the graphic metaphor of Eddington:

> We have found that where science has progressed the farthest, the mind has but regained from nature that which the mind has put into nature. We have found a strange footprint on the shores of the unknown. We have devised profound theories, one after another, to account for its origin. At last, we have succeeded in reconstructing the creature that made the footprint. And lo! it is our own.[73]

There is a further consideration bearing on the uncertainty principle which deserves comment. It is often asserted that the element of ambiguity which quantum physics has exposed in the fundamental operations of nature, whatever its significance may be within this miniature world, does not affect our observation of events in the "molar" or macroscopic world of everyday affairs.

The conventional procedures of classical mechanics, in this view, remain as accurate as ever for "large-scale" happenings in which vast numbers of particles are involved.[74] In a limited and strictly practical sense this supposition is perhaps unobjectionable; but if it is taken to mean that the structure of Newtonian law embracing the visible world has remained unaffected by the quantum revolution, it is seriously misleading. For while it is true that for ordinary purposes the traditional procedures are still "pragmatically" valid —i.e., adequate for the measurements which we normally undertake—it is no less true that the principles which formerly supported them are in shambles.[75] The new and very different laws which have been discovered in the realm of quanta do not "stop at the atom's edge"; if they may be generally disregarded on the macrophysical level it is not because this territory is beyond their jurisdiction but only because our sensory and mechanical equipment is too gross to detect their presence. The "quanta are however there," as de Broglie has put it, "and their existence entails in principle all the consequences we have enumerated. . . ."[76]

What is no longer "there" is the mechanistic assumption of clockwork certainty which underlay the Newtonian cosmology and was so aptly embodied in Laplace's demon. Where once it could be assumed as an article of faith that perfect precision and infallible prediction were in principle within the power of science —and, accordingly, that any inexactness in our findings was only a temporary and technical impediment—today it is all but universally acknowledged that the data of classical physics are at best approximations ("limiting cases" of the quantum theory) doomed forever to an irreducible imprecision not merely in practice but in principle as well.[77] The difference which this makes in the outlook of science, and perhaps not only of science, is considerable. The dogmatic determinism of the scientific mechanist was as rigorous as it was all-embracing; it countenanced no loose talk of capriciousness or chance in any corner of the universe.[78] "The determinist," said Eddington, "is not content with a law which ordains that, given reasonable luck, the fire will warm me; he agrees that that is the probable event, but adds that somewhere at the base of physics there are other laws which ordain just what the fire will do to me, luck or no luck."[79] But Eddington went on to maintain that the old "primary or deterministic law" of classical physics had permanently receded before the new "secondary or statistical law" which leads only to probabilities.[80]

Indeed, although the principle of uncertainty has reference to very small particles and events, it does not at all follow, as Bronowski has seen, that these events are of small importance. "They are just the sorts of events which go on in the nerves and the brain and in the giant molecules which determine the qualities we inherit. And sometimes the odd small events add up to a fantastic large one."[81] Two such fantastic large events, in which very appreciable consequences are triggered by disturbances in one or a few atoms, have been identified by Heisenberg and Jordan; namely, the mutation of genes in hereditary processes, and (more acutely to the point) the construction of an atomic bomb, where we can specify only the upper and lower limits of explosive power but cannot make exact calculations of strength in advance.[82] In this connection it is worth recalling the words of another prominent physicist, written nearly a decade before the advent of the thermonuclear weapon, that "in a domain so alive, so full of future possibilities, as that of atomic and nuclear physics, quanta do play an essential role and it is totally impossible to interpret the phenomena without appeal to them."[83]

Complementarity: The Tolerance of Ambiguity

I can well think that without Bohr's interest and genius we would not today have a broad theory of complementarity in our culture. . . . We would know how to analyze observation and measurement in the atomic domain. But we would not have this immense evocative analogy to situations of psychological and human experience that Bohr has given us.
—J. Robert Oppenheimer, "The Growth of Science and the Structure of Culture" (1956)

The time has come to realize that an interpretation of the universe—even a positivist one—remains unsatisfying unless it covers the interior as well as the exterior of things; mind as well as matter. The true physics is that which will, one day, achieve the inclusion of man in his wholeness in a coherent picture of the world.

—Pierre Teilhard de Chardin,
The Phenomenon of Man (1959).[84]

Heisenberg's uncertainty relations, as we have seen, demonstrated the impossibility of simultaneously measuring such "canonically conjugate" quantities as the position and velocity of an atomic particle—although either one could at a given moment be described with precision. This paradoxical situation found a curious parallel, during the early phase of the quantum revolution, in the inability of physicists to choose between or reconcile the two prevailing theories of the ultimate nature of matter: those of the "wave" and the "particle."[85] The somewhat older particle (or corpuscular) theory served well to explain a wide range of phenomena, but broke down in the face of further tests. The wave theory, on the other hand, while it accounted perfectly for facts resistant to the particle explanation, failed to make sense of those which the older theory had accommodated. In short, both formulations persisted in remaining valid for some observations, but invalid for others; each failed where the other was successful.[86]

It was Niels Bohr who furnished the solution to this crisis through his principle of complementarity, which made it possible to accept *both* theories as valid—not simultaneously but in alternation.[87] The two concepts of waves and particles were said to be "complementary," meaning that both were required for a complete explanation but that they were mutually exclusive if applied at the same time. In effect the two pictures of reality could not exist simultaneously; they could never conflict with one another because they could never meet. "We are continually expecting a battle between the wave and the corpuscle: it never occurs because there is never but one adversary present. . . . They are like the two faces of an object that never can be seen at the same time but which must be visualized in turn, however, in order to describe the object completely."[88]

Another, and perhaps more nearly literal, application of complementarity is to be seen in the mutually exclusive character of the measurements of position and velocity of atomic particles, as described above; in this case the measurements themselves are regarded as complementary.[89] Whichever example is used, the point of Bohr's principle is its recognition that either of the alternatives is partial and inadequate by itself; that, in fact, it represents an "idealization" which is essentially artificial until supplemented by the other of the two factors. A striking illustration of this point has been cited by de Broglie in the relationship of an individual physical unit, such as an electron, with the system in which it has its being. To seek to describe the individual entity with exactness is, so to speak, to sever it from its world; but this forcible isolation cannot be accomplished without a "mutilation of the individuality" of the unit; for the system in quantum physics "is a kind of organism, within whose unity the elementary constituent units are almost reabsorbed." The dilemma which this poses for the investigator, and the conceptual compromise it necessitates, reveal the new and unaccustomed *tolerance of ambiguity* that characterizes the perspective of quantum physics: "The particle cannot be observed so long as it forms part of the system, and the system is impaired once the particle has been identified."[90] These ideal conceptions of unit and system, de Broglie points out, are indeed related to "reality" and useful for describing it, *but only on condition that we do not pry too closely;* for "if we insist on perfectly exact definitions and, at the same moment, on a completely detailed study of the phenomena, we find that these two notions are idealizations, the probability of whose physical realization is nil."[91]

Bohr's principle of complementarity has been described by one of his colleagues as the culmination of the modern philosophy of science,[92] and by another as the clue which unravels the entire domain of atomic experience.[93] As these high tributes suggest, the concept has had reverberations far beyond its original reference to the problem of quantum uncertainty. Thus, among other things, a general theory of predictions for physical systems has been constructed around it, which contrasts complementarity with the causality principle of classical physics in terms of the difference between "subjective" and "objective" theories.[94] A variety of implications for logic and epistemology have also emerged, among which is the construction of a multivalued "logic of complementarity" departing sharply from the conventions of Aristotelian tradition.[95]

Of more immediate significance for the sciences of man, however, are certain still broader intimations first drawn by Bohr and subsequently elaborated by various of his co-workers. From the outset, the principle was recognized as one of potential humanistic reference. "The fact that in an exact science like physics," wrote Max Born, "there are mutually exclusive and complementary situations which cannot be described by the same concepts, but need two kinds of expression, must have an influence, and I think a welcome influence, on other fields of human activity and thought."[96] Inspired largely by this premonition, a substantial body of physical scientists have once again turned their attention to the possibilities of extrapolation—or, more modestly, of analogy—from their "exact" inorganic science to the more complex and confused sciences of life and man. But the disparity between their approaches to this question and those of their Newtonian predecessors is scarcely less conspicuous than the theoretical differences between classical and quantum mechanics. It is not too much to say that the prevalent attitude among leading physicists today toward the human uses of physical techniques is diametrically opposed to the aggressive pretensions of the classical period. Far from pressing for the systematic reduction of the materials of life and society to the methods and mechanisms of traditional science, the most authoritative spokesmen for the new physics are instead to be found insisting upon the high degree of *inappropriateness*, not to say irrelevance, of these methods—and hence upon the essential limitations of physical analysis when carried beyond the borders of its original and proper domain.[97]

They have done so, moreover, not by way of mere *obiter dicta* but, to an impressive degree, after independent investigation of the distinctive materials of the life sciences. Thus Bohr, himself the son of a distinguished physiologist, came to be scarcely less concerned with the analogies and implications of quantum theory for the study of life than with its inorganic matrix.[98] Much the same may be said for Jordan, whose extensive investigations of the biological effects of radiation have led him to a "quantum-physical" (i.e., indeterminist) interpretation of elementary life processes; for Weizsäcker, whose *History of Nature* is an informed survey of the relevance of physical science for the understanding of human nature; for Michael Polanyi, whose *Personal Knowledge* represents a magisterial effort to rescue the human sciences from the dead hand of mechanical philosophy; and in varying degree for Heisenberg, Born and others whose professional careers have lain within the exact sciences.[99]

The application of Bohr's principle to the study of man directly expresses this awareness by modern physicists of the partial and restricted character of their method. For the crucial meaning of complementarity in the context of human affairs—the "immense evocative analogy" it points to—is that of the mutually antagonistic but peculiarly cognate relationship between the traditional scientific method of "causes and mechanisms" and the traditional humanistic method of purposes and reasons—the method known to social science as *verstehen*.[100] It is not denied that the subject matter of biology and psychology *can* be submitted to rigorous investigation in physicochemical terms; what is in question is the adequacy and fruitfulness, even the propriety, of such mechanical analysis.[101] This pervasive sense of limitation with regard to the causal reduction of living systems is illuminated by the observation of Bohr that before such objective probing can be carried quite far enough to tell us exactly what we wish to know, the life which we have under observation is likely to have expired.[102]

To those behavioral mechanists for whom the purposes of science in any field are exhausted by the twin operations of prediction and control, this relentless surgery must seem the only possible procedure.[103] But to others who—like the philosophers of an earlier age and the quantum physicists of our own—contend that the aim of any human science is finally to *understand* rather than to manipulate its subject matter, there is a clear alternative. The nature of that alternative has been well-described by Michael Polanyi:

> The most important pair of mutually exclusive approaches to the same situation is formed by the alternative interpretations of human affairs in terms of causes and reasons. You can try to represent human actions completely in terms of their natural causes. . . . If you carry this out and regard the actions of men, including the expression of their convictions, wholly as a set of responses to a given set of stimuli, then you obliterate any grounds on which the justification of those actions or convictions could be given or disputed. You can interpret, for example, this essay in terms of the *causes* which have determined my action of writing it down or you may ask for my *reasons* for saying what I say. But the two approaches—in terms of causes and reasons —mutually exclude each other.[104]

The inference from this example is not, of course, that the systematic search for natural causes and coefficient correlations must be abandoned forthwith in human affairs, nor that explanation in the qualitative terms of reason and free will is alone sufficient to account for all behavior. The point is rather that the two alternative perspectives or frames of reference are *complementary*: i.e., mutually exclusive if applied simultaneously but mutually "tolerant" if considered as opposite sides of the same coin—differing faces of the same reality.

What is most pertinent of all is the explicit recognition, by a distinguished body of experimental and mathematical physicists, that rigorous physical and mathematical descriptions may be something less than complete or universally applicable in the study of human behavior. Put another way, the most remarkable thing about the principle of complementarity in its analogy to human affairs is not so much *what* it says (the insight has doubtless been anticipated by several millennia of philosophers and seers)[105] as *who* is saying it. What is extraordinary is the fact that it should have found its origin within the most exact department of the most exact of natural sciences, and that it is accordingly the product, not of poetic inspiration or mystical rapture, but of sober scientific observation and experiment. If there is little novelty or force left today in the familiar strictures which have always been leveled against the scientistic spirit by romantics and religionists, there is startling originality in a modern scientific doctrine which deliberately stresses, not the power and glory, but the finite boundaries and ultimate insufficiency of its own method of comprehending reality.

The name of "uncertainty" given by Heisenberg to his quantum principle was, it may be, more appropriate than even its author at first suspected. For it expressed something more than the discovery of a residuum of chance in nature; it gave recognition as well to a note of incorrigible ambiguity in the human act of obser vation itself—in the "argument," as Heisenberg was to put it, "between nature and man in which science plays only a part." It is curious indeed that many who are now prepared to concede the full impact of this discovery in quantum physics should continue emphatically to deny its suggestive relevance to human concerns. If there is a core of ambiguity in the relations of man with the comparative "still life" of inorganic nature, there would seem to be prima facie grounds for allowing the possibility of an equivalent ambiguity, no less permanent and irreducible, in the relations of man with man—in the human encounter of an "I" and a "Thou." For a problem of "uncertainty" or "indeterminacy" arises, in whatever field of knowledge, from the presence of a degree of complexity and subtlety which renders the subject matter recalcitrant to the keenest apparatus (whether mental or mechanical) that can be brought to bear in its analysis. Until it has been demonstrated that human behavior is less complicated in its variability and contingency than the behavior of microphysical systems—that it possesses, in Conant's term, a "lower degree of empiricism"—it would seem only elementary scientific caution to hold similar reservations with regard to its rigorous causal analysis.[106]

More than a few scientists, in fact, have been prepared to press this "evocative analogy" a good deal farther—as we shall see in some detail in the chapter to follow. The extent to which these speculations have been carried may be suggested here by a single illustration. Recalling the "idealizations" involved in Bohr's concept of microphysical complementarity, the French quantum physicist de Broglie proceeded some years ago to a generalization which it would be conservative to call unorthodox. May it not be universally the case, he asked, that the concepts produced by the human mind are "roughly valid for Reality" so long as they are formulated in slightly vague fashion—but that "when extreme precision is aimed at, they become ideal forms whose real content tends to vanish away?"[107] Citing as an example the simple notion of an "honest man," which finds its counterpart in the real world only when somewhat vaguely defined ("like the plane monochromatic wave, absolute virtue, if defined with too exacting a precision, is an idealization the probability of whose full realization tends to vanish"), de Broglie arrived at this very un-Cartesian proposition: "Thus in the region of the inexact sciences of human conduct, the strictness of the definitions varies inversely as their applicability to the world of Reality."[108]

De Broglie's speculation depended equally upon the principles of indeterminacy and complementarity. Whenever we set out to describe facts, in any field of observation, we soon find ourselves dealing on the one hand "with a Reality which is always infinitely complex and full of an infinity of shades, and on the other with our understanding, which forms concepts which are always more or less rigid and abstract." The effort to come to grips with this elusive reality, to make our understanding correspond to the facts, is commendable and necessary. But what is very much in doubt is "whether such a correspondence can be maintained to the end, if we insist on eliminating the margins of indeterminateness and on effecting extreme precision in our concepts." Emphasizing that even in the most exact of the natural sciences the need for "margins of indeterminateness" has repeatedly become manifest, de Broglie left no doubt of his own conviction that in the notoriously inexact sciences of human nature and conduct those indeterminate margins must, if anything, be still more generously and loosely drawn. His conclusion was that a complementary relationship obtains in all realms of knowledge between what he termed (following Pascal) the "spirit of geometry," aiming always at precision, and the "spirit of intuition," which "recalls to us without ceasing that reality is too fluid and too rich to be contained in its entirety within the strict and abstract framework of our [scientific] ideas."[109]

The degree to which this unconventional reflection of a physicist has come to find support in the "postmodern" sciences of life and man will be explored in subsequent chapters. For the moment it may be enough to underscore the broadest and most reliable inference for the humane sciences which has thus far emerged from the physical revolution of our century: that as the mechanistic viewpoint has been found to be inadequate for the full comprehension of inorganic matter and natural events, it is a fortiori inadequate for the understanding of human nature and human events; and, more specifically, that the assumption of objective predeterminism upon which all consistent causal analysis (with its corollaries of exact prediction and control) must finally depend is, in simple fact, without confirmation in the new physics of possibility.

This was the general conclusion arrived at, in the early 1930's, by the renowned physicist Arthur H. Compton.[110] However, since Compton went on (no doubt ingenuously) to express certain personal convictions concerning the existence of God, his modest inference from physical theory was widely dismissed as irresponsible obscurantism. Particularly harsh in their rejection of Compton's inference were certain adherents of the rigorous philosophy of science known as logical positivism (whose ancestral ties to the behaviorists we have already noted). For this reason, if for no other, it is rewarding to recall the conclusion which was reached in the same period by one of the most authoritative spokesmen for the positivist philosophy, Hans Reichenbach:

> It is of crucial importance that the solid barrier which determinism erects against every non-deterministic solution of the problem of life and freedom has fallen, that we can no longer speak of objective predetermination of the future, and that the concept of *possibility* and of *becoming* takes on an entirely new aspect when we no longer need regard it as an illusion due to human ignorance, as a mere substitute for the description of real and objectively existing facts, which are only subjectively withheld from us human beings.[111]

CHAPTER V

The Ambiguity of Life:
The Biology of Freedom

> Owing to this essential feature of complementarity, the concept of purpose which is foreign to mechanical analysis finds a certain field of application in biology.
>
> —Niels Bohr

IN ORDER TO FIND LIVING ANALOGIES to Heisenberg's discovery of indeterminacy in nature and Bohr's principle of its toleration, it is not necessary to resort at once to the notoriously ambiguous sciences of man and politics. The reverberations of these discoveries are hardly less evident in the "neutral" zone of biology, lying intermediate between the physical and social sciences. It is true, of course, that biology has always displayed inveterate if furtive tendencies toward vitalistic and teleological explanations. "Teleology," remarked von Brücke in the last century, "is a lady without whom no biologist can live; yet he is ashamed to show himself in public with her."[1] In the very heyday of the scientistic spirit, when it appeared that all of life must soon be vivisected into its discrete and palpable determinants—even after Darwin's genius had supplied the organic missing link to Newtonian mechanics—there were still those who stubbornly insisted upon the qualitative differences which common sense perceived in the

singular life processes of growth, organization, regulation and direction—in a word, of purpose. And yet, as a philosopher of science
has recalled, "biology was unable to keep away from the triumphal
progress of the causal idea," whose steady advance had penetrated
every corner of organic activity until it had explained "such
phenomena as digestion, respiration, and heart's activity—yes, even
the activity of the brain—as physico-chemical processes."[2]

Later still, after the initial momentum of the Darwinian
revolution had subsided and the ingenuity of physicochemical explanations seemed exhausted, even then nonmechanical hypotheses,
while they grew more numerous, were generally met with disdain
—or worse, with indifference. "It is self-evident," wrote Max
Verworn in the *Encyclopedia Brittanica*'s eleventh edition, ". . .
that only such laws as govern the material world will be found
governing vital phenomena—the laws, i.e., . . . of mechanics."[3] So
long as biologists waited for their cues at the door of the physics
laboratory—and while the only sound to be heard from within
was the clank of mechanical models—this assumption was indeed
"self-evident." Life scientists felt themselves committed at the
outset to the supposition, as Reichenbach has put it, "that physiological processes must ultimately be reducible to mechanical motions of atoms and corpuscles, and that, accordingly, all the
imperfection of causal explanations which we observe can be only
provisional."[4]

Today these formidable assumptions of the materialist faith
retain little of their power to intimidate the biologist and inhibit his
exploration of new pathways in his own science.[5] Classical causation, in the rigorous sense of reductionism and determinism, is no
longer triumphant in physics; the model of the great machine has
gone to join the unlamented Aether in the oblivion of conceptual
anachronisms; the imperfection of causal explanations has turned
out to be not provisional but inherent and inescapable; and, most
decisive of all, exact science itself has pressed the case for the
limitations of science—or, at least, of that conception of science
and its method which passed as definitive until only yesterday (and
still circulates unchallenged in the tidy milieux of behavioral sci-

ence). "The time of materialism is over," writes Max Born with the courage of the new conviction: "We are convinced that the physico-chemical aspect is not in the least sufficient to represent the facts of life, to say nothing of the facts of mind."[6]

The Microphysics of Life

The contribution of physics to the life sciences, moreover, has not consisted solely in the critical and preliminary feat—immense as that has been—of sweeping aside the dead hand of mechanism. Various of the root metaphors and guiding concepts of present-day biology have in fact found their sources in the anticipations of physical scientist–philosophers.[7] Of these the most fundamental is undoubtedly the principle of "wholeness," which implies nothing less than the irreducibility of physical systems from the standpoint of their full scientific explanation. This is broadly the implication of Heisenberg's uncertainty principle, further developed by that of complementarity; more concretely, it finds expression in de Broglie's description (cited earlier) of the indissoluble relationship between the electron and the system of which it is a part. Thus, as the biologist von Bertalanffy has summarized, "a principle of wholeness appears in microphysics"—expressed in the fact that the elementary events cannot be further broken down but "can be treated only as a whole."[8]

Closely related to the principle of holism is the conception of *organization* or "organism," conveying a recognition of pattern and form, of internal order and regularity, as inherent qualities of every system in nature. For a famous example, Whitehead's philosophy of organism visualized the atoms of physics as minute "organisms" differing in size but not in essential quality from the living matter studied by biology.[9] More recently, the physicist-philosopher L. L. Whyte has argued for a unitary principle in physics and biology which emphasizes process, development and transformation, and "recognizes at the start a *universal formative process* in nature, a process in which regular spatial forms (symmetrical patterns) are developed and transformed."[10]

Whatever may be thought of the particular merits of these theories of universal process and organism in nature—not to mention those of "panpsychism," which literally imports mind into the physical universe[11]—it is clear that they have found inspiration in the conceptual breakthrough of quantum physics. "Wholeness, organization, dynamics," writes von Bertalanffy: "these general conceptions may be stated as characteristics of the modern, as opposed to the mechanistic, world-view of physics."[12]

Some physicists, indeed, have gone beyond analogy to propose the existence of actual conditions within living systems directly parallel to those which produce the uncertainty relations in micro-physical systems—conditions, that is, which result both from the "atomic-physical fineness" of their reactions and from the disturbance which necessarily accompanies their investigation. William G. Pollard in particular has maintained that, if this hypothesis holds, "those properties of the [biological] system connected by indeterminacy relations would, as in the case of physical systems, lead to the contrasting phenomena involving them which Bohr designates as complementary."[13] Noting that the disturbance produced by the observation of living organisms is everywhere more substantial and pervasive than in physical science, Pollard suggests that an impassable minimum may exist "below which the disturbance produced in the [biological] system by the carrying out of any such procedure could not be reduced by any further ingenuity." His conclusion, documented by impressive testimony from biologists, is that mechanico-mathematical forms of analysis need to be "complemented" in the study of life by alternative methods of understanding which take into account the unique, irreducible and unquantifiable characteristics of the live subject matter.[14]

The still bolder contention that living organisms are directly governed in significant aspects of their behavior by the laws of quantum mechanics—and hence are not to be classed with inorganic macrophysical structures supposedly exempt from those laws—has been advanced by the physicist Pascual Jordan, whose major contributions to the development of the quantum theory give unusual weight to his views. In contrast to large-scale nonliving systems, consisting of countless atoms all of the same sort and subject to the same external conditions, living systems are held by Jordan to be radically different in their composition and complexity, exhibiting a fineness of structure which extends below the level of microscopic analysis to colloidal and molecular dimensions. "Correspondingly, the quantities of matter which take part in certain very fine but decisively important physiological reactions apparently often embrace only a few molecules."[15] On the basis of specific biological and physiological examples, Jordan conjectures that the ultimate relations governing organic life "are of absolutely atomic-physical fineness." But this finding does not, of course, support an inference that the deterministic laws of classical mechanics are applicable to these ruling conditions of life. On the contrary, it leads to the conclusion "that the organism is quite different from a machine and that its living reactions possess an element of fundamental incalculability and unpredictability."[16]

Like Pollard, Jordan holds that there are definite limits to the possible refinement of measuring instruments and hence to any reduction of the interference brought about by observation—that "here also a fundamental complementary relation exists, which appears to be a characteristic of the living." In this version of Bohr's analogy, the freedom of living things is a freedom not *without* but *within* the law—a freedom made possible because the unusual law which governs these vital relations is better described as a statute of self-limitation, carrying its own built-in interstices.[17]

The Recovery of Purpose

It should be noted, however, that few present-day biologists, even of the most "holistic" persuasion, are able to find much nourishment for their own science in such direct extrapolations from the statistical laws of microphysics.[18] The really decisive impact of the quantum revolution upon the life sciences has been more indirect and analogical; it has thrown open heuristic doors and admitted invigorating drafts of fresh air into the laboratory. For if there was nothing completely novel to biology about the concepts of wholeness, organization and process, they received from the physicists a powerful new sanction and respectability. As Harold G. Wolff has written:

> The revolt in physics against the Cartesian concept of a mechanical universe raised doubts about the ideal model for science imposed by physics. Far from being disrupting, this change made it easier for many biologists to admit into the study of the form and function of parts of living systems their purpose in relation to the goals of the living organisms and to accept the thesis that biological concepts can emerge from a study of integrated systems in which new and different relations between creature and setting engender new and different behavior patterns.[19]

The physical eclipse of mechanism, in short, has emboldened biologists to approach their own subject matter in synthetic as well as analytic terms—to perceive *Gestalten* where once they could see only additive assemblages of parts. And with this radical shift in perspective has come a sharpened awareness of the meaning of *process* in living systems: namely, the recognition of organisms as active and dynamic centers of directive striving rather than as inert or static mechanisms responding only passively to external stimuli. In a word, the concept of *purpose* was restored to biology, shorn of mysticism but not of teleology; living organisms could now be understood as *acting upon* the world (not merely being acted upon) in pursuit of goals set from within. Thus von Bertalanffy has distilled the essential principles of the organismic theory in three succinct phrases: "The conception of the system as a whole as opposed to the analytical and summative points of view; the dynamic conception as opposed to the static and machine-theoretical conceptions; the consideration of the organism as a primary activity as opposed to the conception of its primary reactivity."[20]

The recovery of purpose in biology, however tenuous and contested in various corners of the field, would seem by now to be firmly established.[21] When biologists came to observe the functional relations of parts of a living system to the whole, as they contemplated the wider relations of these "open systems" to their environment, and as their vocabulary assimilated notions of formative process and dynamic activity, "the intrusion of purpose could be avoided only by those biologists who abandoned biology."[22] Nor can this resurgence of purpose be simply dismissed as a retreat into the ingenuous vitalism of an earlier period—which admittedly contributed little more than convenient labels (whether "entelechy" or *élan vital*) for all that orthodox theory left unexplained. The modern conceptions of goal-seeking and directive activity are no less solidly grounded in systematic observation and experiment than their physicochemical counterparts. "Such teleology," Sinnott has written, "far from being unscientific, is implicit in the very nature of organism. The biologist need not shudder at these words, for purposiveness of this sort is not only unobjectionable in his science but lies at the very heart of life itself."[23] Variously defined and developed within such conceptual frames as "organicism," "holism," "purposivism," "telism," "mentation," "goal-directiveness," etc., the recognition of purpose is now so pervasive and routine in biological literature as to seem secure even from the attacks of logical positivists and psychological behaviorists.

Furthermore, although the present-day representation of purpose is distinct from earlier vitalistic theories, it is not an automatically disqualifying defect that natural science should on this occasion come to terms with common sense—and that ancient intuitions of purposiveness once smuggled in through the back door should now be welcomed openly at the front. We have seen that the insight involved in this common-sense view could never be completely suppressed even at the high noon of the mechanistic era. Thus, while the reductive methods of classical physics still appeared omnipotent, Whitehead could be heard protesting that

> ... you cannot limit a problem by reason of a method of attack. The problem is to understand the operations of an animal body. There is clear evidence that certain operations of certain animal bodies depend upon the foresight of an end and the purpose to attain it. It is no solution of the problem to ignore this evidence because other operations have been explained in terms of physical and chemical laws. The existence of a problem is not even acknowledged. It is vehemently denied.[24]

The similar expostulation of J. W. N. Sullivan is perhaps still more in point. Again and again, he maintained, we come to feel that the primary concepts of biologists are inadequate in the face of their most important problems. "The great theory of natural selection, for example, . . . does not in the least explain the most obvious fact about the whole process, that is, the upward tendency of living things."[25]

This "obvious fact" of the upward tendency of living things—which nevertheless no amount of measurement and dissection could apprehend—is no longer universally regarded as biological heresy. More and more life scientists have been forced to conclude that vital processes can never be effectively reduced to physicochemical categories, not because of the imperfection of our measuring instruments but because such methods *reverse the proper procedure in the study of life.* "Actually, vital processes face the other way about, forward not backward, and the key problems of biology are not analytic but synthetic."[26]

The far-reaching implications of this apparently simple idea may be observed in the recent rise of "open system" theories in biology, which defy if they do not overthrow the second law of thermodynamics in its application to life. That law holds that the entropy of any physical system—roughly equivalent to its degradation of energy—is always on the increase (with the ultimate result, in Eddington's famous extrapolation, that the largest physical system of all, the universe, is inexorably running down). But it now seems likely that, as one theorist has concluded, "The entropy content of a living system is a completely meaningless notion."[27] For only if physical systems are entirely "closed" (isolated) can it be said that entropy always increases in them. Living things, however, are evidently never perfectly isolated, never in true equilibrium or static contentment; they are "open" to the world and engage in regular exchange of materials and energy with it. The evolution of life itself, in the words of Julian Huxley, "is an anti-entropic process, running counter to the second law of thermodynamics with its degradation of energy and its tendency to uniformity. With the aid of the sun's energy, biological evolution marches uphill, producing increased variety and higher degrees of organization."[28]

In this "open-ended" approach, the essential characteristic of life is not process merely but *tendency;* not a settling but a striving; not permanence but change. Its normal direction is toward ever-increasing levels of complexity and heterogeneity, with accompanying tension and effort—rather than toward homogeneity, the reduction of tension and the restoration of equilibrium at a lower level. The struggle of life is not for survival but for growth; its goal is not being but becoming.

Among the more instructive insights which have contributed to theories of organicism and holism is that of "equipotentiality" (or "equifinality"). It was Hans Driesch, one of the last of the "old-fashioned" vitalists, who first demonstrated the prodigious regenerative capacity of organisms. He discovered that the embryo of a sea-urchin, when divided in half, grew not into two unfinished demi-creatures (as mechanistic theory would have it) but rather into two complete and normal organisms. Numerous other experiments have since confirmed that living things at various levels display a remarkable ability, when confronted with crippling obstacles, to discover or "invent" new ways of reaching their original goals. Unlike machines, which can be put through the paces of their artificial cycle only on the basis of fixed initial conditions, living organisms show an inner resourcefulness and ingenuity which enable them to fulfill their purposes under drastically altered circumstances. This "equipotential" phenomenon— the sheer marvel of morphogenetic regulation—is, to be sure, rarely interpreted nowadays as the manifestation of an inward entelechy or demon piloting the organism. On the other hand, it has remained stubbornly impervious to all attempts at explanation in physicochemical terms.[29] Something of the puzzle which equipotentiality presents—as well as the imaginative leap toward understanding that numbers of scientists have felt obliged to take —is suggested by a notable pronouncement of the renowned embryologist Hans Spemann. Observing that he had been drawn again and again to use terms which point not to physical but to psychical or mental analogies, he went on to assert:

> This was meant to be more than a poetical metaphor. It was meant to express my conviction that the suitable reaction of a germ fragment, endowed with the most diverse potencies, in an embryonic "field," its behavior in a definite "situation," is not a common chemical reaction, but that these processes of development, like all vital processes, are comparable, in the way they are connected, to nothing we know in such a degree as to those vital processes of which we have the most intimate knowledge, viz. the psychical ones. It was to express my opinion that, even laying aside all philosophical conclusions, merely for the interest of exact research, *we ought not to miss the chance given to us by our position between the two worlds.* Here and there this intuition is dawning at present. On the way to the high new goal I hope to have made a few steps with these experiments.[30]

In the same manner, Michael Polanyi has become convinced that the processes of equipotentiality constitute "a primordial form of originality" or spontaneous inventiveness, attesting to the presence of an active center in animal life which cannot be specified or comprehended in the mechanical terms of physics and chemistry. In effect the classical method, while it can *analyze* minutely whatever is placed before it, cannot ever *recognize* living forms in themselves: "Physical and chemical knowledge can form part of biology only in its bearing *on previously established biological shapes and functions*: a complete physical and chemical topography of a frog would tell us nothing about it as a frog, unless we knew it previously as a frog."[31] Our recognition of the frog as a frog (the Whole Frog, or frog-in-itself) is independent of and antecedent to whatever notations we may make about it as detached observers; and, in turn, our awareness of the frog qua frog recedes and vanishes in the process of its reductive analysis. And so it is, a fortiori, with observation at other and higher levels of the animal kingdom.[32]

It would be difficult to imagine any more striking biological parallel to the principle of complementarity than this. At first glance indeed there would seem to be *two* parallels, two separate forms in which the principle finds expression in the realm of life. One statement emphasizes the complementarity of *our knowledge* of organisms; the other stresses the complementarity of organic processes themselves. Polanyi's monumental study (as its title, *Personal Knowledge,* indicates) represents an exhaustively researched brief on behalf of the first of these expressions: i.e., that scientific knowledge embraces the complementary inquiries of "causes and reasons"; that the so-called subjective values of personal commitment and passionate participation are as integral to science as rigorous analysis and mathematical abstraction. Our observation of life, in short, requires alternately the dispassionate probe of the measuring instrument and the empathic communication—the shock of recognition—of one living form encountering another.[33]

The second expression of biological complementarity consists in the awareness of an "unspecifiable" (i.e., irreducible and unmeasurable) quality in living things, an active source of rudimentary mentation, existing side by side with the physiological mechanisms which yield themselves to measurement and reduction. But it is immediately apparent that this formulation is only a slightly different way of expressing the same principle. *Both the actual processes of organic life and our knowledge about them operate on two complementary levels.* One level is that of *mechanism*—mechanical in process and mechanistic (physicochemical) in analysis. The other level is *free from mechanism,* both in process and in observation. On either description, then—that of organic processes or of our knowledge about them—mechanism is only half the story, in the deepest sense a half-truth. The other part of the story is that which biologists are driving at, or toward, in those imaginative and boldly unconventional conceptions which previous pages have surveyed. "Here and there this intuition is dawning," said Hans Spemann in the passage quoted above—adding that the way to the "high new goal" may hopefully have been lighted by his own researches into those vital processes of organisms which are like nothing so much as the processes of our own minds. Another biologist of prominence, the late Ralph S. Lillie, has stated the case more directly: "The general conclusion to which we are led by these considerations is that in living organisms physical integration and psychical integration represent two aspects, corresponding to two mutually complementary sets of factors, of one and the same fundamental biological process."[34]

A further speculation which has pertinence to our theme has been contributed by the brilliant expositor of the organismic theory, von Bertalanffy. Observing that the laws of science in their application to successive levels of nature comprise "a hierarchy of statistics," von Bertalanffy declares that in this hierarchy "we find a remarkable phenomenon which we can describe as *an increase in freedom*."[35] The significance of this is not only that as we move upward in the hierarchy our laws more and more take on the character of gross statistical averages, approximately as "deterministic" as actuarial tables; more important, there occurs an increasing variability and contingency of individual development in terms of form, of structural arrangements, even of chemical processes ("Here too, with increasing complexity we find an increase in the degree of freedom"). Most pertinent of all, as we have already seen, there emerges a widening range of alternative (equipotential) courses of action open to the resourceful selection of the individual.

> Thus, in the hierarchy of statistics the degree of freedom seems to increase as we move to higher levels; not in the sense of the indeterminacy of elementary physical events but rather in the sense that the process as a whole follows definite laws, *but different possibilities are left open to the individual events.*[36]

There is accordingly a sense in which it has become scientifically appropriate—no longer imprudent or impertinent—to speak of the *freedom* of living organisms. Indeed there would seem to be more than one way in which the elements of contingency and opportunity enter into the careers of individual organisms. But the significant freedom of living things is that of an active volitional center endowed with a degree of spontaneity or primitive creativity which defies analysis in terms of mechanical causation. The organism, as Pi Suner has put it, "is a unity having spontaneity. . . . The chance, the need for variation in biology, is something which does not apply to physics; hence it seems imperative to acknowledge in a biological phenomenon some variable and spontaneous element which would not appear through physical causes."[37]

This rudimentary resourcefulness on the part of organisms is, however, not at all impervious to human observation and recognition; we know of it precisely by virtue of our "position between the two worlds." That unique vantage-point enables us to take the role of the living form, to penetrate the veil and enter into its perspective—*because we ourselves are forms of life*.[38] The specter of "anthropomorphic subjectivism," which has haunted the behaviorists of every field, thus appears to be not a fatal corruption but an invaluable human prerogative—an essential component of scientific knowing.[39] What the new biology tells us is simply that if we wish to understand the nature and behavior of any living thing, we are perforce compelled to take an *inside* as well as an *outside* view; we cannot longer decline to recognize, with Teilhard de Chardin, that "coextensive with their Without, there is a Within to things."[40] And, when we have gone this far, we may even find it possible to share the conviction of the same great scientist of life

> . . . that the two points of view require to be brought into union, and that they soon will unite in a kind of phenomenology or generalized physics in which the internal aspect of things as well as the external aspect of the world will be taken into account. Otherwise . . . it is impossible to cover the totality of the cosmic phenomena by one coherent explanation such as science must try to construct.[41]

The phenomena of life—if not the whole of physical reality—can be grasped in their fullness only by the willing suspension of scientific disbelief. That much, perhaps, has been said before by nonscientists. But what is newly significant is that it is no longer scientific *belief* which we are required to suspend.

Humane Evolution: Possibility and Choice

A century after the appearance of *The Origin of Species*, it seems ironically appropriate that the doctrine of human evolution should have come to be described and accounted for in terms of "descent"—rather than of "ascent." For the image of human nature which Darwin's theory projected forced man to descend even farther, and to shrink still more, in his scientific estimation of himself than the low estate to which he was already accustomed. "Epic of science though it is," as Loren Eiseley has said of Darwin's book, "it was a great blow to man. Earlier man had seen his world displaced from the center of space; he had seen the Empyrean heaven vanish to be replaced by a void filled only with the wandering dust of worlds. . . . Finally, now he was to be taught that his trail ran backward until in some lost era it faded into the night-world of the beast."[42]

This is, however, no longer the ruling perspective in which Darwin's heirs, the evolutionary biologists of today, contemplate their subject. In effect that perspective has been completely reversed: although they are not inattentive to the darkening trail of the past, what fascinates the neo-Darwinists of our time is its uncharted course into the virgin future. Where formerly the emphasis was on the *origin* of all species, today the interest of evolutionists is focused upon the *destiny* of one in particular.[43]

There has indeed been something like a softening of the general outlines of Darwinism over the past half-century, even while its central features have been conclusively confirmed. Most obviously of all, the ethics of "tooth and claw," which had extrapolated the jungle images of competitive struggle to human and social development, have faded to furtive echoes in the journals of political paranoia. (A poignant testimony to their passing may be seen in the shocked reception accorded Sir Arthur Keith's *Evolution and Ethics* when it appeared in the 1940's; that last work of a great scientist, an almost perfect recapitulation of the rugged theories of Spencer and T. H. Huxley, was greeted as a startling and rather ludicrous anachronism.)[44] On other levels, meanwhile—owing largely to the persistent and persuasive work of Ashley Montagu—the process of organic evolution has come to be seen as characterized at least as much by cooperative endeavor as by competitive struggle; while the "struggle" itself is now perceived in terms not merely of survival but of growth and the fulfillment of possibilities.[45]

But by far the greatest change which has come over Darwin's theory is not so much a correction as an addition. Organic evolution, with all of its paraphernalia of mechanism and blind push, has not terminated abruptly with the advent of man; beyond doubt it goes on as inexorably as ever, not only in the jungle and sea around us but in the marrow of man himself. But it is no longer the prime mover, no longer really significant; it has, quite simply, been superseded and surpassed. Something new has been added—or, if not precisely new, something so qualitatively different in scope and action as to constitute a transformation. In man, to paraphrase a remark of Julian Huxley, the evolutionary mechanism has become conscious of itself—*and so ceased to be mechanism*. The rigorous laws of physical and phylogenetic determinism have been not merely transgressed but suspended; henceforth, as V. Gordon Childe has succinctly put it, "Man Makes Himself."[46]

This momentous transformation has occurred through the agency of a wholly new kind of evolution—a process not organic or mechanical but intellectual and social. In technical terms (which recall Lamarck rather than Darwin), the new evolution has been described as "the inheritance of acquired social characters."[47] More simply, it is the transmission of experience through culture. Its distinctive reality is as much a scientific fact of observation as that of the older organic evolution, and no less profound in its implications (if much less swift in its popular impact and recognition). The most evident of these implications is also the deepest: *the new evolution rests not on mechanism but on freedom*. The very concept is itself an objective correlate of human freedom and a testament to its existence.

This hypothesis of what might be called "humane evolution" is now so broadly accepted in the literature as to be virtually an integral part of the modern synthetic theory of evolution. Among its protagonists are such names as Julian Huxley, C. H. Waddington, T. Dobzhansky, E. W. Sinnott, P. B. Medawar and Chauncey Leake.[48] But no one has put the case for "the new evolution" more forcefully than the distinguished paleontologist George Gaylord Simpson, whose explanation is worth following at some length.

> In organic evolution new factors arise as mutations without volition and without any fixed equivalence to needs or desires. Once they have arisen, their fate or further role in evolution is determinate and follows from the interplay of factors in which the organisms assist but over which they have no semblance of control. *In the new evolution new factors arise as elements in consciousness,* although always somehow influenced by and sometimes directly produced from the tangled psychological undergrowths of habit, emotion, and the subconscious. They arise in relationship to needs and desires of individuals and commonly in relationship to the individual's perception and judgments of the needs and desires of a social group. Once they have arisen, *their further evolutionary role is not mechanistically determinate* and is subjected to the influence not only of the actual needs and desires of the group and of volitions extremely complex in basis but also of an even more complex interplay of emotions, value judgments, and moral and ethical decisions.
>
> Through this very basic distinction between the old evolution and the new, the new evolution becomes subject to conscious control. Man, alone among all organisms, knows that he evolves and he alone is capable of directing his own evolution. For him evolution is no longer something that happens to the organism regardless but something in which the organism may and must take an active hand.[49]

"Man, then," as another writer has summarized, "is undergoing a two-phase evolution, the older organic and the newer social evolution."[50] In this two-phased process is to be seen, more vividly than ever, yet another and higher expression of that "immense evocative analogy" of complementarity which we have found applicable to other levels of reality. To be sure, evolution in its organic phase is not, on our previous argument, strictly mechanical; even simple organisms betray the symptoms of purpose and the rudiments of mentation. But if there is a plausible case for freedom in the subhuman world, it is prima facie more convincing and conclusive in the world of man. The complementarity of human evolution in its dual phases—one of them blind (or near-blind) and the other volitional and intentional, one a thing of determinate *causes* and the other a matter of self-determined *reasons*—is only one line of evidence among many in defense of the case for human freedom. But it is the evidence of incontrovertibly expert witnesses, consensually validated on highest scientific authority—and supported, if more humbly, by the persistent testimony of common sense.

The new way of evolution, moreover, is not only rooted in freedom; it also aspires toward freedom as its goal.[51] Human evolution is no longer merely pushed from behind by a blind force; it is actively drawn by a vision. It is curious to reflect that evolutionary biology, which had done so much only a short time ago to dethrone and humble man in his pride, should now agitate to restore him to his kingdom with newly impressive credentials of legitimation.[52] But this is perhaps the mark of its own theoretical evolvement. For if the prevailing synthesis of thought does not look upon modern man as the final completed goal of evolution (and which of us would be content with that conclusion?), it is firm in its view of him as the sole plausible trustee of the evolutionary process—literally, the last best hope of earth.[53]

More than that: it regards man as capable of fulfilling this responsibility. Few scientists of any field are more acutely aware than evolutionists of the crises and abysses that yawn before humanity on its immense journey; for they have, after all, been professionally charting the course. They know the perils of over-population and underadaptation; the danger of predators without, and of the barbarian within. More accurately than any other group of specialists, they can project the future probabilities on the basis of calculations covering roughly a million years. And yet, for the most part, they will give you odds. . . . "It is another unique quality of man," writes Simpson at the conclusion of *The Meaning of Evolution*, "that he, for the first time in the history of life, has increasing power to choose his course and to influence his own future evolution. The possibility of choice can be shown to exist. This makes rational the hope that choice may sometime lead to what is good and right for man."[54]

With the expression of this contingent faith in mankind—couched in terms of possibility, never of certainty—it may be protested that such theorists of evolution have gone beyond the legitimate boundaries of their science. But that would be to deny them competence at the very heart of their subject. The most significant and substantial of their data are no longer just the predictable certainties and determinate causes of mechanistic analysis. Their observations also encompass a fruitful finding of "indeterminacy," a biological residuum of inescapable choice which is as much a fact of life (in its human stage if not in all others) as are the complementary processes of measurable mechanism. Their belief in human freedom is no mere preferential whimsy—although it surely confirms a preference—but a scientific conviction imposed and validated by the living stuff of observation.

Consciousness: The Vital Emergent

Both of the complementary phases of evolution traced by science—the organic and the humane—are processes pointed toward the future. Whatever their other differences, in either case this purposive direction entails increasing organization, heterogeneity, complexity—and ingenuity. In the undoubted fact of upward succession to ever-higher levels of subtlety and order lie, as we have seen, the origins of freedom.[55] Organic freedom appears to observation first in the form of an elementary and ephemeral resourcefulness, barely sufficient to cope with trivial accident or primitive calamity: enough, in short, to permit survival and growth. But at each successive stage of existence the increment of freedom to which von Bertalanffy has drawn our attention becomes more insistently and pervasively apparent. The higher the order of life, and the greater the challenge presented to it, the more resourceful and creative is its response.

It is not difficult to see where this must lead us. There is no need to invent a novel term (entelechy, telefinalism, *élan vital*) for this subliminal capacity to deal with crisis and opportunity that living organisms display with mounting decisiveness at each step of the ladder. The designations for it are old and well-established; but they are terms which scientists are loath to admit into their vocabulary, for reasons sufficiently familiar to us. Those terms are, of course, *mind* and *consciousness*. Merely to mention these pariah words in scientific discourse is to risk immediate loss of attention and audience. Yet if there is anything of validity in the new directions and discernments of biology with which this chapter has been concerned, the most fundamental phenomena of life—and the course of organic evolution itself—cannot be comprehended otherwise than in some such terms as those of evolving mental activity and emerging consciousness. This unorthodox viewpoint requires, of course, the redefinition of mind not as a static and completed structure but rather as an activity progressively unfolding and developing—not the noun-form "mind" but the verb-form "minding" (to borrow Herrick's term).[56] The rudimentary freedom of animal organisms—cribbed and cabined as it is by phylogenetic blind alleys and dead ends of development—is nothing else than a freedom of crude mentation, a germination of consciousness and choice. The goal of organic evolution, the purpose of life, is to bear this vital seed of tendency to fruition and so to transcend and renew itself at the next higher level: to bring about in each ascending cycle a new birth of freedom and a widened glimmer of consciousness.

And, in its turn, the goal of the new "psychosocial" evolution of man—which presupposes consciousness and has long since reached self-consciousness—is to fulfill the higher promise of its own unique responsibility and singular striving: not the merely precocious mastery and manipulation of *things,* but the fully mature understanding of the *self* and of the *other*.

That is the task pre-eminently conferred upon the sciences of man.

CHAPTER VI

The Freedom To Be Human:
New Directions in Psychology

In the latter half of the nineteenth century, the influence of biology and its related disciplines upon the new field of psychology was broadly and aggressively "mechanistic": that is, analytic and reductive. But in the first half of the present century, as we have seen, the impact of the science of life upon the study of mind has been almost as decisively "organismic": that is, holistic and constructive.

It was only to be expected, no doubt, that as biologists came to recognize in their subject matter the characteristics of purposeful striving and resourceful spontaneity—to perceive the animal organism as a whole and ongoing concern—they should seek to extend their findings to human organisms as well. Perhaps it was also to be foreseen that they would come into conflict once again, on this new and higher ground, with the mechanistic adversary who had long since established beachheads in most corners of the field. It is not entirely whimsical, indeed, to regard some of the warmest controversies of present-day psychology—in such differing arenas as those of personality, perception, even psychoanalysis—as local engagements in a larger battle between mechanists and holists over the mind of man. Moreover, in this figurative conflict it is surely the mechanists who have sought to subjugate the human mind to "external powers"—and the nonmechanists who have sought to set it free.

The Evidence of Psychobiology

The freedom of the mind—in the double sense of its autonomy from mechanism as well as of its power to choose goals and carry out purposes—is an inference directly supported by the work of those pioneer life scientists who have dared to press their researches beyond conventional boundaries into the wilderness of mental experience. For a classic example, the dean of American organismic biology, G. E. Coghill, was led by his experiments in embryology not only to the principle of purposive striving but to the ultimate ascription to the human animal of intrinsic powers of creativity and self-direction. While he was not quite prepared to claim that a secure basis had yet been laid "for a scientific conception of such a personal freedom as seems necessary for the existence of personal moral responsibility," Coghill argued forcefully that there was no longer any scientific ground for a doctrine of "determinism or fatalism." He concluded that the only rational approach to the problem of freedom was through the hypothesis of the human person as primarily a system of self-initiated action rather than of reaction—as the creator no less than the creature of his organic destiny.[1]

In these conclusions, as in the studies that underlay them, Coghill and his colleague C. Judson Herrick were following a path first charted by the latter's older brother, Clarence L. Herrick —an early explorer of the specialized terrain which came to be known as psychobiology.[2] It is remarkable that the rigorous experimental researches of these organic scientists, spanning three-quarters of a century in time, did nothing at all to encourage the supposition that psychological processes might somehow and someday be reductively accounted for in physicochemical terms. On the contrary, the psychobiologists were more and more strengthened in their conviction that the operations of the human mind are inherently impervious to translation or conversion into the material categories of physical science.

Thus, in the volume which represents the summary and climax of his own life's work (*The Evolution of Human Nature*), the late C. J. Herrick gave renewed emphasis to his conclusion that "awareness cannot be 'reduced' to physico-chemical categories or adequately described scientifically by ever so complete an explanation of the mechanism employed. . . . The most characteristic feature of mental action is that it can be known only introspectively."[3] In this belief he and Coghill were long ago joined by another great psychobiologist, Adolf Meyer, whose psychiatric psychology made room not only for "common sense" but for the method of introspection as an indispensable tool of psychological inquiry.[4]

The extent to which this once-dreaded specter of introspection (with its explicit acknowledgement of the "within of things") has gained scientific standing through psychobiology may be seen from Herrick's definition of the field as "the scientific study of the *experience* of living bodies, its method of operation, the apparatus employed, and its significance as vital process, *all from the standpoint of the individual having the experience.*" While the physiologist may undertake to observe these activities as a spectator from the outside, the psychobiologist is constrained to "view them from the inside." Far from banishing his own personal experience as a contaminating bias, he actively invokes it as a source of illumination. And the means of understanding which such a scientist finds primarily appropriate and fruitful are those of seeking "to put himself in the place of the person or animal having the experience, to see the situation from the other's point of view."[5]

But this is, of course, no casual undertaking. To seek entry into the inner experience of another human being—to propose to gain an *intimate* understanding of his thoughts and wishes—is, in psychology as elsewhere in life, to initiate a personal encounter which carries with it certain formidable consequences of commitment and responsibility. They are, broadly, the consequences that follow from *respect for the human subject*[6]—which means taking him seriously "on his own terms," both in his conscious purposes and in the deeper values he is seeking to actualize.[7]

The principal interest of psychology, then, is in the experiencing person himself, both in his wholeness and his uniqueness. While remaining scientific, that interest is genuinely a personal interest—not "impersonal" or indifferent. By entering into the perspective of the other individual, in seeking to know his private intentions and needs, psychology can no longer regard him as a *means* (whether tool, statistic, experimental datum or machine) but comes to see him for himself, as an *end*: in Mumford's words, as "the Whole Man, man in person." The object of investigation, at the level of psychology, has become a subject; and science's concern *with* him is henceforth inseparable from its concern *for* him. In adopting his standpoint and taking his role it has had to take him seriously—with his values, his intentions and his reasons. Implicit in the recognition of this personal engagement is a new acceptance of responsibility, of a degree and kind unfamiliar to the traditional canons of science. But it is a responsibility not at all unfamiliar to the greatest of scientists. Perhaps its finest expression is to be found in the concluding words of a classic work of modern science, *Man on His Nature*, by Sir Charles Sherrington: "Ours is a situation which transforms the human spirit's task, almost beyond recognition, to one of loftier responsibility. . . . We have, because human, an inalienable prerogative of responsibility which we cannot devolve, no, not as once was thought, even upon the stars. We can share it only with each other."[8]

It is especially appropriate to our present theme that it should have been Sherrington—the dean of British physiology and possibly the most esteemed life scientist of the present century— who gave voice to this eloquent affirmation of humane responsibility. For no one yet has presented the biological case for the essential human freedom—the freedom of the mind—more convincingly than Sir Charles, in his Gifford Lectures published as *Man on His Nature*. After a lifetime of research on the nervous system, in particular on the brain and its mechanism,[9] Sherrington had no message of comfort for the mechanists of psychology. For all he knew, he told them, the human mind could never be reduced to physical energy nor comprehended in its terms: "The two for all I can do remain refractorily apart. They seem to me disparate; not mutually convertible; untranslatable the one into the other."[10] If mind were in reality a form of energy it should be measurable quantitatively, by resort to the energy-scheme.[11] But the search for such a scale arrives at none; the human mind stubbornly resists all efforts to take its measure, and shrinks forever from the objective probe of the analyst.[12]

But if there is no equivalence of the physical and mental, if we are compelled to admit that "energy and mind are phenomena of two categories," how is this vexing duality to be tolerated and these polarities reconciled? Sherrington's answer constitutes a vivid intuition of the principle of complementarity as applied directly to the human mind: "While accepting duality we remember that Nature in instance after instance dealing with this duality treats it as a unity."[13] It has evolved us as compounds of energy and mind, joined together in an underlying and overriding wholeness —"a great graduated scale of seething organization," whose epic achievement to date has been the production of mind and the values of minding.[14]

The wholeness of man, like the organization of evolving earth itself, rests not on the identity but on the *harmony* of these two distinctive components of mind and energy. To seek to melt that unity down by a naive monism, whether spiritual or material, was for Sherrington "an irrational blow at the solidarity of the individual; it seems aimed against that very harmony which unites the concepts as sister-concepts. It severs them and drives off one of them, lonelily enough, on a flight into the rainbow's end."[15] The human passion for a more convenient and less ambiguous order, one comprehensible to our senses and amenable to our instruments, only disturbs and distorts the harmony of that true order. ("That," said Sherrington, "is the old urge in all of us, intent on relating phenomena, and for that purpose finding 'causes.'") The quest for measurable certainty has indeed proved triumphantly successful elsewhere, making the energy-concept a weapon for man's conquest of the earth. But it has had no success with the "I-thinking, its ways, and its creations." The progress of our knowledge, and especially of natural science, has only made more evident the essential powerlessness of these spatial concepts to deal with or describe the mind. "Mind, for anything perception can compass, goes therefore in our spatial world more ghostly than a ghost."[16]

Yet that insubstantial mind has given us all that counts in life: "Desire, zest, truth, love, knowledge, 'values' . . ." We know its reality by direct experience of these of its effects and affects. How then is it related to the energetic and substantial world, the world of dimension and duration? Sherrington's answer speaks not only of complementarity but of priority:

> They have this in common—we have already recognized it—they are both of them parts of one mind. They are thus distinguished, but are not sundered. Nature in evolving us makes them two parts of the knowledge of one mind and that one mind our own. We are the tie between them. Perhaps we exist for that.[17]

The two realms of human experience, and the two ways of knowing—objective and subjective, mechanical and volitional, measurable Without and unspecifiable Within—are each demonstrably valid and real enough to suit the standards of this exacting physiologist. Yet they are not separate-but-equal: "they are both of them parts of one mind." One of them, only one, *comprehends* the other—in the twofold sense of understanding and embracing.[18] Mind may take the measure of the substantial world, but cannot be measured by it or in its terms. Mental experience, as Herrick also reminded us, is part of the vital process of human life. As such it is, indeed, "natural"—natural to man, but preternatural to mechanism. A naturalism which knows only what it can measure knows nothing at all of this experience. But man reflecting on his nature—i.e., man *introspecting*—knows better.

Perception: A Paradigm of Behavior

Few corners of the intricately partitioned field of psychology have been more vigorously mined and cultivated over the past half-century than that of perception. More and more this preliminary feature of human behavior—the act of noticing—has come to be regarded as central to the concerns of other departments and divisions of the field. A measure of its recognition is indicated by the fact that, for all their organized "perceptual defenses" against invaders, the specialists of perception have seen their borders overrun by a succession of aggressive movements (behaviorism, functionalism, Gestalttheorie, psychoanalysis) reflecting the particular interests of such diverse and alien powers as personality, learning, motivation, clinical psychology and social psychology.[19]

The first, and still the most durable, influence upon perceptual theory was that of the mechanistic-elementarist school which found its modern expression in behaviorism. Within the rigorously objective framework of the S-R scheme, students of perception were led to focus their attention upon the measurable attributes of the outer *stimulus*, on the one hand, and upon the "autochthonous" or structural elements in perceptual *response*, on the other.[20] We have already seen (in Chapter II) something of the origins and career of this ambitious attempt to locate the springs of all human action in mechanical and physicochemical determinants. But in addition to its modern scientific appeal the behaviorist approach was consistent with more venerable assumptions concerning the dichotomy of man and nature—more exactly, of the receptive organism and the impinging world. Because human beings in ordinary experience appeared all but powerless in the face of external forces, it seemed plausible to give attention to the overriding impact of the environment upon man, rather than to the puny actions of man upon the environment. Things and events "out there" were the controlling factors, and human behavior the passive and predictable response.[21] Prometheus might seek to rebel against the fates, but in the end (as someone has said of the heroes of Hemingway) he was only a man to whom things were done.

It is notable that the first substantial reaction of the present century to this behaviorist approach—that which became known as the "Gestalt revolution"—was not revolutionary enough to alter the predominantly structural, objective and mechanical orientation of perceptual theory.[22] In the Gestaltist framework, to be sure, the configurational field of forces playing upon the individual became much more complicated and highly organized than in the linear equation of S-R. To the forces emanating from the stimulus-object were added other forces of a neural and physiological character within the responding subject (specifically within the brain); and perception then became in effect the end product of their interaction.[23] There can be little doubt that the Gestaltists, in their special application of the "holistic" principle to perception as well as in their apprehension of the phenomenal field, were the agents of a constructive advance carrying with it fresh and fruitful insights. But among their insights was not that of an active perceiving individual whose perceptions are related to his present personality system (self or ego), his past experience (learning), and his future-directed purposes (motivation). Perceptual behavior on the classical Gestalt model was still the determinate outcome of *forces* over which the individual had no control and little comprehension— "forces in the object," to use Hamlyn's words, "interacting with forces in the cortex in a specifically causal way. Perception is treated as the end-product of a process of stimulation, *not something in which we engage*."[24]

The invasion of Gestalt was, however, not the last assault upon the ramparts. After its contributions had been assimilated, together with those of the behaviorists, it appeared that everything was accounted for in perception but the *perceiver*. It was this conspicuous gap which the subsequent "New Look" movement—the invasion of functionalism—set out more or less deliberately to close. In so doing, it not only reacted against the prevailing structuralism of research but transformed the whole field of perceptual psychology.

The fundamental innovation of the new look was a radically altered perspective in which the act of perception came to be seen as a selective and dynamic function of the total personality, an ongoing process or "transaction" through which the individual creatively relates himself to his world. Since the first expressions of this new approach began to appear in the 1940's, an accumulation of experiments and observations have confirmed what common sense might seem to testify: namely, that perception is not an inert and indiscriminate response to irresistible forces, but a highly discriminative process in which, as it were, many stimuli may be called but few are chosen.[25] The concept of selective perception also implies a corollary, that which the late Harry Stack Sullivan labeled "selective inattention" and which others have called "perceptual defense"—i.e., the capacity of the individual to ignore, reject, transform, and generally reconstruct the materials of his experience.[26]

The implications of the functionalist viewpoint for perceptual psychology have been far-reaching. Its preliminary principles of selectivity and resistance imply the presence of both cognition and intrinsic motivation—not merely trial-and-error learning but the coherent *organization* of experience by the perceiver. A variety of attempts have been made over the past score of years to erect conceptual schemes to account for this evident fact of perceptual organization. Sherif, for one, taking perception as the prototype of all experience, posits "a central structuring or patterning" of perceptions by the individual—made up of such factors as "motives (needs), attitudes, emotions, general state of the organism, effects of past experience, etc."[27] Postman and Bruner have identified the subjective organizing principle as a set of "hypotheses"— that is, "expectancies or predispositions" which serve to select, organize, and even transform the stimulus information crowding in from the environment.[28] The ultimate purpose of these students has been to develop "a unified theory of behavior" which will treat the individual as an organized whole and regard his perceiving as "an instrument of adjustive activity."[29]

Other researchers in recent years have corroborated the general conclusion of these functionalists that such personal motives as needs, values and purposes enter significantly into the process of perception. "Value as a variable of personality" has been tested and affirmed by one group; the determining role of past experience has been demonstrated by others; even the "need for achievement" has emerged as a factor in the recognition-time of word association tests.[30] Finally, investigators with a psychoanalytic or personality-centered approach have emphasized the influence of emotional strivings and subjective wishes upon perceptual behavior.[31]

This explicit recognition of the perceiver as an active agent in the course of his own perceiving is, on its face, a substantial departure from the structural and mechanical approaches which formerly held the field. But it must be admitted that the full implications of the functionalist new look have not been accepted with equal enthusiasm by all students of perception. In particular, those with commitments to behaviorism have sought to reduce the undeniable influences of motivation and cognition to the more manageable status of "intervening variables."[32] They have preferred to redefine the selective element in perception in terms of more or less blind drives and attitudes lying latent within the individual and only awaiting a stimulus for their release. By this method it has apparently been hoped to make room for the perceiver while yet retaining the classical emphasis upon the objective "stimulus situation." But the perceiver, once admitted into this "black box," is effectively shut out again; he is regarded not as the active chooser and interpreter of his perceptual experience but as a passive, not to say neutral, organism in whom attitudinal sets are "sprung" or dammed-up drives triggered off at the touch of an external stimulus. His behavior thus comes to be explained fundamentally as a kind of commerce on the reflex level between organic drives or impulses and impinging forces from the environment, with the human subject as little more than the field for this "psychophysical" interplay.

But this is far from the full significance which most functionalists have thought to find in their researches. The very term "functionalism" itself would seem to carry a more definite and

positive meaning—at least that meaning which has traditionally attached to the older and broader school of the same name in general
psychology. Ever since its origins some three-quarters of a century
ago in the writings of James, Dewey and Angell, functionalism has
conveyed the pragmatic view of psychological processes as instrumental agencies helping man to cope with his environment.[33]
Viewed in this way, the functionalist attitude has never been
wholly absent from American psychology during the intervening
decades, and is markedly characteristic of its namesake in present-
day perceptual study. Moreover, of the many variants of functionalism today the most direct lineal descendant of Dewey's theory is
the viewpoint known as "transactionalism"—a vigorous current of
contemporary thought and experiment which takes its name and
inspiration straight from the master.[34]

The preliminary groundwork for transactionalism and related theories of perception was first laid in a classic article by
Dewey on the reflex-arc concept, published in 1896.[35] This short
paper contained the outline of the entire philosophy of human
behavior subsequently elaborated by Dewey, G. H. Mead and
others; and it furnished, incidentally, an anticipatory rebuttal to
S-R "behaviorism" several years before the behaviorist movement
was formally inaugurated. Dewey's article was a concise and positive statement of the creative and volitional, as well as social, character of human behavior (including perceptual behavior). The
central point was that the "stimulus" to behavioral action is never
a simple sensory excitation but rather is the state of the total organism, the unique coordination of ongoing processes which constitutes a personal perspective. It is this to which an individual
"reacts"—or, more accurately, toward which he *acts*. Behavior for
Dewey was therefore a selective, seriated and coordinate action:
"The stimulus [as he wrote later] is simply the earlier part of the
total coordinated serial behavior and the response the latter part."[36]
This conception implies what its author called the "correlativity of
stimulus and response," in which attention or activity first determines the stimulus, and the stimulus in turn determines further
activity—a principle which subsequently led both Dewey and

Mead to a theory of the mutual determination of form (organism) and environment. What is significant for our purposes is the thesis that the individual is in the fullest sense an *actor* (not a mere field or receptacle), who creatively determines his environment by selecting and reconstructing the materials of experience in terms of his own sensitivities and makeup—his unique personal perspective. "What the individual is," in Mead's words, "determines what the character of his environment will be. . . . Man sets the universe out there as like himself, identical in matter and substance."[37]

These expressions are scarcely to be distinguished from those voiced by numerous perceptual theorists today, on the basis of empirical evidence unavailable in Dewey's time. Thus a leading member of the transactional school defines perception as "that part of the process of living by which each one of us from his own particular point of view creates for himself the world in which he has his life's experiences and through which he strives to gain his satisfactions."[38] Our perceptions, in this contemporary view, are crucial to all our subsequent behavior; they provide us, to use Adelbert Ames's phrase, with "prognostic directives for action." Moreover, the process of perceiving is itself active and creative; it is not a mere registering of sense-data but a volitional act of selection, a *choice*, among all the available opportunities of the perceptual field. Whether wholly or only partly conscious, the choice rests with the perceiver, operating from his unique behavioral center. Perception in any human instance, in other words, can only be understood in "first-person" terms; it cannot be adequately represented or comprehended without reference to the *personal intentions of the perceiver*—what he is trying to do or to become.[39] In Cantril's succinct phrase, "There is no *at*tention without *in*tention."[40]

Once again we are brought to recognize what the evidence of psychobiology (if not of all biology) has already demonstrated: that there can be no genuine understanding of behavior without an understanding of the *behaver—on his own terms*.[41] No objective analysis of discrete elements in perception, artificially abstracted from the experienced process—and no amount of detached observation of *Gestalten* from a neutral perspective—can tell us what the perceiving individual means to do in his world, and therefore what that world means to him.

But the perceiving person himself *can* tell us. One of the most remarkable developments in contemporary perceptual theory has been what may quite simply be called the rediscovery of consciousness. What is especially in point is the discovery of numerous investigators that the subjects of their tests are able to report their perceptions, to talk about their own experience, with unsuspected accuracy and reliability. "Throughout the literature on perceptual learning," write two prominent theorists, "we have found innumerable studies in which subjects were queried as to what they were aware of during the studies, and time and again it has been found that results have co-varied with verbalized reports of awareness."[42] In short, *the subjects have known what they were doing.* The same authors proceed, in a chapter on "The Place of Consciousness and Conscious Meaning in Perception," to make a confession which is, to say the least, unusual in its candor: "Throughout this book we have tried to stay at a simple observational level. This has been grossly dissatisfying; the operations and their results merely inform us that something is happening and we are left in the dark as to what this is."[43] In other words, so long as they have remained at the objective observational level, without taking the perceiving subject into account (and into confidence), the investigators quite literally have not known what was happening. Their own (properly cautious) conclusion is "that consciousness of what is going on in a perceptual learning experiment is an extremely important variable."[44]

Paralleling this rediscovery and reinstatement of the conscious subject in perception is a recent trend in the neighboring field of motivational theory—the dawning recognition, in Gordon W. Allport's words, "that the best way to discover what a person is trying to do is to *ask* him."[45] The conventional procedures of investigation in this area, despite the fact that the knowledge they seek is of personal motives, effectually disregard the subject's own conscious report as untrustworthy in favor of probing into "hidden causes," whether genetic or physiological, of which the subject is (by definition) unaware. At no point, as Allport observes, do these methods ask the subject what he is up to, what he wants or what he is trying to do; as a result, *the individual loses his right to be believed.*[46] But, in marked opposition to these prevailing assumptions, the findings of numerous depth-probing tests in recent years have strongly confirmed the individual's right to be believed by demonstrating that normal persons relate by the "direct method" precisely what they reveal by the projective method. In short, however shocking it may sound, we can take their motivational statements at face value—"for even if you probe, you will not find anything substantially different."[47]

Moreover, not only has the "face-value" method proved accurate with ordinary people in eliciting the truth about personal motives; in various experimental situations it has turned out to be the *only* means of getting at the truth, succeeding where other and more devious methods have failed. After citing one such test, in which a formidable battery of psychodiagnostic devices failed to detect certain motives which were known to be present to the point of obsession, Allport has concluded: "Here is a finding of grave significance. *The most urgent, the most absorbing motive in life failed completely to reveal itself by indirect methods. It was, however, entirely accessible to conscious report.*"[48]

The inference from these developments in perception and motivation would seem clearly to be, not only that the individual has a "right to be believed," but that the observing scientist has an obligation *in the interest of science* to respect the subject's reasons and his reasoning—to give due regard to his version of the truth. ("Some day," wrote the author of *The Organization Man* only a few years ago, "someone is going to create a stir by proposing a radical new tool for the study of people. It will be called the face-value technique. It will be based on the premise that people often do what they do for the reasons they think they do.")[49]

There is, in addition, still another frustration confronting the would-be objective student of perception: a built-in limitation evocative of the "disturbance" factor in the uncertainty relations of quantum physics, and reminiscent of Bohr's dictum that in science, as everywhere in life, man is at once an actor and a spectator.[50] For it is not only the "naively" perceiving human subject who molds his perceptions, and therewith his world, in accordance with his personal behavioral center; the scientific observer-perceiver himself does so no less unavoidably. "Even the scientist who studies perceiving," write Cantril and Bumstead, "is himself perceiving; he enters the transaction as a creative participant, not as a detached, wholly passive, 'objective' spectator."[51] Thus the traditional dichotomy of the observer and the observed —the formal separation of subject and object already cast in jeopardy by quantum theory and further undermined by the biology of freedom—is dissolved altogether at the very threshold of behavioral inquiry: in the preliminary study of perception itself.[52]

At a minimum, then, it would seem that the student of perception must come to terms with the process of his own perceiving, must learn to "know himself" on the introspective level of his private values and intentions. Constrained by the demands of science to embrace his own standpoint, he cannot remain a stranger to himself. On the basis of that perceptive self-awareness he may thereafter come to know and share—to encounter in personal transaction—the perceptual experience of another.

If the uncertainty principle of microphysics (with its lesson of the limitations of objectivity) has found so close a parallel in the study of perception, it might be supposed that the corollary principle of complementarity has relevance here as well. And, indeed, the application of that "vast evocative analogy" is not far to seek. The complementarity of human perception is apparent in the distinction between what Ernest G. Schachtel has called "the two basic perceptual modes": one which is essentially passive and unreflective (although it may include conscious awareness), and another which is characterized by active reflection and acute self-consciousness.[53] The fundamental disparity between these two modes or levels of human perception has been repeatedly remarked by investigators from a variety of traditions dating back at least as far as Leibniz.[54] What their various formulations share in common is the recognition that perception of the first and simpler order is characteristically indiscriminate and receptive, involving little more than the registering of sense-data; while perception of the second order is creative, self-consciously reflective and broadly intentional. In brief, the first perceptual mode corresponds to the elementary reflexive behavior of lower organisms and of infants, whereas the second mode invokes that which is distinctive of human beings in their maturity: namely, the capacity of symbolic reflection and decision, the knowing "which knows that it knows, as well as what it knows."

Solley and Murphy, in their discussion of the role of consciousness in perception, have graphically described the dimensions and boundaries of the first of these two perceptual modes. Following such theorists as Sartre and Wilhelm Stern, they write:

> This level of consciousness, this nonreflective level, is passive, seemingly nonenergetic, and has the characteristics of intensity, clarity, or vividness. . . . Essentially, the only kind of meaning that occurs at this level is the meaning of sensory qualities. . . . *If man were to be conscious at this level only, he would never question the universe, he would never seek to know himself, or seek answers to riddles.*[55]

But man is not conscious only at this rudimentary level; he need not remain arrested at the organic-sensory threshold of immediate experience. His consciousness, in Sartre's words, "can know and know itself." Man in his maturity is not simply receptive to experience but *perceptive toward it.* Through his ability to make indications to himself—to control his own responses as well as to select his stimuli—"man is no longer a slave of the past or of the present, but becomes a master of the future."[56]

Once again, in the psychology of perception, we encounter the underlying complementarity of behavioral processes: the "split levels" of the *organic* and the *humane.* Both of these perceptual levels are realities of experience, verifiable empirically and demonstrable logically. But they are also (to adapt the language of Sherrington) incommensurable, not mutually translatable or reducible one to the other. The two separate perceptual modes require two separate conceptual modes of explanation, corresponding

to their qualitative difference. For perception at the *organic* level—
the level of sheer sensory impression—mechanistic analysis along
the familiar causal and objective lines is doubtless appropriate and
adequate. For perception at the *humane* level—where the process is
a creative synthesis of intention-attention-selection, hence indeterminate and immeasurable—the mode of explanation must be accordingly synthetic and sympathetic. The scientist "may be a *student of*
any other subject-matter. He is always a *participant in* perception."[57] The meaning which he seeks will unfold itself only
through the transactional dialogue of two perceiving subjects

The human study of perception may accordingly be viewed
as a paradigm of the several sciences of man. Through recognition
of their special opportunities as well as of their limitations, perceptual psychologists have been brought increasingly to the realization
that the ultimate goal of their inquiry *as scientists* is neither predictive certainty nor manipulative control, but rather the appreciative
understanding of the experience of another which arises from the
meeting of minds—from the human encounter of an *I* and a *Thou*.
Expressed in such untechnical terms, this conclusion may seem to
reflect a spirit altogether alien to "science." But if the emphatic
disavowal of the mechanistic model in favor of a *person-centered*
one which we have traced in perceptual theory still strikes many
psychologists as a counsel of despair, there is a growing company of
others for whom it represents both a needed affirmation of humane
responsibility and a belated accession of scientific modesty. On the
last page of a notable recent textbook in the field of personality,
one such psychologist (Harold Grier McCurdy) takes leave of his
readers with this boldly unconventional prescription:

> If such a conception of personality seems to give us few
> handles by which to manipulate others, I do not think that is an
> argument against it. Under any conception of personality, other
> persons are anything but easy to manipulate. I nevertheless think
> that we can be helpful to one another. When our friend is in
> despair, we can ask him what it is that he hates. When he is
> happy, we can ask him what it is that he loves. These questions,
> or others like them, may enable him to unfold his personal uni-
> verse in reasonable discourse for his enlightenment and our own.[58]

It may be that there is yet another lesson to be learned from the maturing psychology of perception. We have seen that the search for ulterior causes and behavioral "handles" is coming to be recognized, in this and other fields, as inadequate on scientific grounds. But its irrelevance in terms of science is less important, surely, than its irreverence in terms of humanity. The ultimate question, as an eminent physicist has suggested, is "not whether one *can* experiment with man, but whether one wants to or *may* experiment with him."[59] When the subject loses his right to be believed, what is forfeited is not an abstract privilege but nothing less than his unique identity as a responsible moral agent, his credentials as a person. To deny or disregard his capacity for reflective perception and creative experience is to withhold not only the civil right of "fair hearing," as it were, but the fundamental human right of self-realization—the opportunity of becoming fully himself.

Schachtel, in his searching analysis of the two basic perceptual modes of relatedness—the "autocentric" (lower) and the "allocentric" (higher)—has called our attention to the inescapable *choice* which men must make between these alternative dimensions of experience. To exist on the perceptual level of autocentricity is to prefer a "closed world," free alike of ambiguity and responsibility—one which resembles nothing so much as "the closed worlds of the animals predetermined by their relatively few needs and their innate organization." Where men's perceptions are dominated by this defensive and intolerant perspective, the potentialities of mind and feeling stagnate and atrophy, leaving the senses dulled and the spirit numbed. "The closed world of this perspective ceases to hold any wonder. Everything has its label, and if one does not know it the experts will tell him."[60]

In contrast, the perceptual mode of allocentricity is an "open encounter of the total person with the world," illuminated by creativity and spontaneity. This is the mode of original artistic and scientific achievement, as indeed of all personal growth and becoming. It is a level of perception which, as human beings, we are free to seek and capable of reaching—if we dare aspire to its heights. In contrast to the animals, declares Schachtel, man is capable of continued growth and development throughout his life if he succeeds in remaining open to the world and capable of allocentric interest. . . . The basic difference between animal and human mental organization, man's openness toward the world, can be increasingly realized only if man retains and develops the allocentric mode of perception, the first appearance of which is the most important step in the development of perception in the growing child.[61]

Men must decide, then, which of these alternative perceptual roads to follow: the organic or the humane. If they heed the insistent counsel of behavioral mechanists (in perception or also where), they will dutifully take the former course—less from conscious reflection, no doubt, than from conditioned response. Viewing themselves in this foreshortened perspective, they may then confirm the hypothesis and fulfill the prophecy. They may come to settle for a perceptual existence on the organic plane of passive nonresistance, asserting nothing and accepting all, stimulus-bound and appetite-centered—and so become in truth predictable, controllable, and docile.

Human beings conditioned to apathy and affluence may well prefer this regressive path of least resistance, with its promise of escape from freedom and an end to striving. But we know at least that it is open to them to choose otherwise: in a word, to *choose themselves*.[62] They can limit their vision to the perceptual level of autocentricity—or they can literally "raise their sights" to the level of allocentricity, the demanding plane of self-activation and self-actualization. In the latter case they will take charge of their own perceptions—undertaking to cultivate their senses, to stretch their horizons and to deepen their perspectives. They will come to regard their perceptual experience not only as a personal concern but as a responsible engagement, a continuously creative (and continually uncertain) encounter with the world.

To live at this higher perceptual level requires a voluntary acceptance of risk, a capacity for wonder, and a sustained tolerance of ambiguity. But all of life, as Justice Holmes reminded us, is an experiment: "Every year if not every day we have to wager our salvation upon some prophecy based upon imperfect knowledge."

And more recently, two uncommonly perceptive students of perception have made the same point no less firmly: "Every perception is an act of creation; every action is an act of faith. . . . And in every occasion of living, perception-in-operation is a never-ending process of prediction in the face of uncertainty for action on the basis of faith."[63]

Freud: Romantic Mechanist

It would be too much to say of the history of psychoanalysis that it presents an unbroken narrative of the liberation of the human mind from the mechanical grip of nineteenth-century scientism. But few impressions emerge more forcefully from a review of this embattled movement of thought than that of its gathering reaction against what Sherrington called "that old urge in all of us" to explain human conduct in objective-causal terms, and so to reduce it to the managcable dimensions of energetic forces working on or through a pliant organism.

Such an anti-mechanistic trend is, to be sure, least conspicuous in the mind and work of the great founder of psychoanalysis. It was remarked in our first chapter that one of Freud's earliest independent undertakings was an abortive "Project for a Scientific Psychology," which was designed expressly "to furnish us with a psychology which shall be a natural science; . . . to represent psychical processes as quantitatively determined states of specifiable material particles and so to make them plain and void of contradictions."[64] Although that straightforward project was soon abandoned, the vision behind it remained alive in Freud's thought not only as a far-off divine event but as an active influence in the shaping of his theoretical and therapeutic premises.[65]

It was not, of course, that Freud set out as a matter of deliberate strategy to win the approval of orthodox scientists for his novel ideas. In his own attitudes toward science, as his great biographer has made clear, Freud was eminently a child of his time, of his culture, and specifically of his intellectual milieu. Ernest

Jones has reminded us that in the Victorian period of Freud's formative years science meant, "not only objectivity, but above all exactitude, measurement, precision."[66] That it meant even more than that to many of its devotees, including Freud himself, has also been emphasized by Jones:

> Moreover, in the nineteenth century the belief in scientific knowledge as the prime solvent of the world's ills—*a belief that Freud retained to the end*—was beginning to displace the hopes that had previously been built on religion, political action, and philosophy in turn.[67]

Like other adolescents before him and since, the youthful Freud "had the need to 'believe in something' and in his case the something was Science with a capital."[68] His will to believe found fulfillment in long years of academic absorption in the natural sciences leading to the medical degree: three years in biology and chemistry, followed by no less than six years of dedicated apprenticeship in Ernst Brücke's laboratory of animal physiology. As a student and disciple Freud was completely "captivated" (to use Jones's term) by the aggressive materialism of Brücke, which asserted that all organisms including the human are nothing but systems of atoms moved by forces obedient to the principle of the conservation of energy—forces ultimately reducible to those of attraction and repulsion. The extent of Freud's debt to Brücke and his school is permanently displayed in the remarkably similar vocabulary and imagery in which he came to formulate the basic concepts of psychoanalysis.[69]

His subsequent associations with such significant others as Breuer, Charcot and Fliess (rigorous organic scientists all) served only to reinforce and deepen Freud's commitment to natural science on both its mechanistic and evangelistic sides. When in the late nineties he came to articulate his general theory of mind and more specifically his interpretation of dreams, the intellectual resources he had to draw upon were almost exclusively those of the natural sciences and their medical correlates. Indeed, the wealth of his sophistication in these fields stands in astonishing contrast to the comparative poverty of his background in psychology itself.[70] It is highly instructive (as well as ironic) that the author of the most influential and comprehensive system in modern psychology should have possessed no formal training and little more reading acquaintance within his own chosen field. It would appear that in his approach to the mind, lacking a key to the conventional front door of consciousness, Freud had no choice but to enter through the basement—with all its darkly impressive tangle of power machinery, naked pipelines and seething cauldrons.[71]

In view of all this, the wonder is that Freud ever reached his destination at all. But the fact is incontrovertible that he did. Despite the extreme physicalistic bias of his personal outlook and conditioning environment—despite (it almost seems) his own overt intention—it was after all the *mind* that came under observation and analysis by Freud: not the brain or the nervous system, nor, least of all, the reflexive responses of an organism to outer stimuli. In various crucial aspects—in its underlying assumptions, its major insights, and most clearly in its method of treatment—Freudian psychoanalysis has always been profoundly (if uncomfortably) at variance with the postulates and techniques of organic medicine on the one hand and of experimental psychology on the other. The concern of the analyst, for example, is inescapably with the patient as a unique and whole person: the two partners in the therapeutic act engage in a dialogue predicated upon their mutual faith in the power of reason to know the truth—and upon the further faith that the truth will make men free. In his theory no less than his therapy (as we shall see later on), the major conceptions of Freud more nearly resemble those of a poet than of a scientific mechanist. Recognizing this, we might well agree with the psychiatrist Karl Stern that all of Freud's important discoveries were first of all "felt from within," won originally through introspection and confirmed by intuition, culminating in theoretical expressions which, for all their freight of scientific metaphor, cannot ever be translated or reduced to the categories of physical science.[72]

It is true, nonetheless, that Freud himself never abandoned the hope that such a reduction would eventually be brought about —despite the incorrigible tendency of his own doctrines in the opposite direction.[73] Unable to divert the main current of the psychoanalytic movement from its "unscientific" course—denied, for that matter, even the evidence of his own clinic for his mechanistic premises—Freud could yet adopt a "metapsychological" posture of rigorous psychogenetic determinism in order hopefully to restore the balance. The declared faith in conscious reason which underlay his whole therapeutic enterprise ("where id was, there shall ego be") did not prevent him from insistently minimizing the role of reason as an actual or potential determinant of personality and conduct—nor, on the other hand, from maximizing the thrust of irrational forces pressing their claims both from "below" (id) and from "above" (superego). In Freud's famous topographical layout of the mind, the ego (itself only partially conscious) never achieves complete autonomy but ekes out a precarious existence as a kind of buffer state between the rival tyrannies of instinct and introjected culture—of animal nature and social nurture.[74] On one side is the "seething cauldron of excitement," welling up with tidal periodicity and threatening to spill forth its molten lava of libido with consequences devastating to all concerned; on the opposite side lurks the sternly repressive and authoritarian agency of *Kultur* itself, incessantly moralizing, policing and proscribing.

Indeed, to Freud in his later philosophic mood all of human history, both ontogenetic and phylogenetic, appeared as the history of psychic struggle—an elemental conflict between opposing inner forces which bear a curious resemblance to the classic antagonists of dialectical materialism (making it plausible to say of Freud that he stood Marx on his head). Thus the id symbolically represents the oppressed "revolutionary" stratum, striving to release its energies in explosive violence and destruction; while the superego presents an image of the spuriously paternalistic ruling authority, shrewdly camouflaging its exploitive interest beneath a cloak of opportune morality.

"Every individual," according to Freud, "is virtually an enemy of culture. . . . Thus culture must be defended against the individual."[75] A major strategy of this cultural defense was that of infiltrating a fifth column (the superego) into the heart of enemy territory, somewhat on the theory of divide-and-conquer. "Civilization thus obtains mastery over the dangerous love of aggression in individuals by enfeebling and disarming it, and setting up an institution within their minds to keep watch over it, like a garrison in a conquered city."[76] The ensuing struggle between the individual and society, as Freud discerned it, was at bottom an expression of the primordial conflict between the instinctual forces of life and death. "And now, it seems to me," he concluded in *Civilization and Its Discontents,*

> the meaning of the evolution of culture is no longer a riddle to us. It must present to us the struggle between Eros and Death, between the instincts of life and the instincts of destruction, as it works itself out in the human species. This struggle is what all life essentially consists of and so the evolution of civilization may be simply described as the struggle of the human species for existence.[77]

It is a curious situation that today, after a generation of post-Freudian exegesis, opinion is still divided over which of the contending forces in this internecine struggle—nature or nurture, id or superego—was the favorite of Freud. Although the division falls broadly along school-of-thought lines, the significant differences are not so much between psychological "liberals" and "conservatives" as among the "liberals" themselves. The conservative critique of Freud (if the strictures of reactionary opponents may be so dignified) has always been as unambiguous as it is irrelevant; those conscientious objectors to his system who view themselves as the guardians of culture and morality (hence of the superego) never seem to recover from their initial shock at what they take to be simply a mischievous brief for sexual license and libertinism, aimed at the very foundations of civilization. This characteristic conditioned response of Philistinism would scarcely be noteworthy were it not for a recent revelation of its underlying animus which compels attention by its sheer outrageousness. Prior to the appearance of David Bakan's carefully documented study portraying Freud as Devil's advocate (*Freud and the Jewish Mystical Tradition*), the sporadic attacks upon psychoanalysis from what might be called the cultural Bible belt shared a hidden premise which could be insinuated but never uttered aloud—being at once libelous and comical. The premise was that Freud must in some mysterious but authentic way have been bedeviled, a dupe of Antichrist, and his whole movement therefore resolved into a summons to damnation. This is essentially the thesis now openly advanced and defended, with a wealth of scholarly detail, by Professor Bakan. Specifically, his contention is that Freud entered into a metaphorical "Satanic Pact" in the years immediately before the turn of the century—a negotiation provoked by a period of creative sterility and personal depression, the fruit of which was the successful establishment of the psychoanalytic movement. Freud, we are told, while he did not literally believe in the existence of the Devil, clearly recognized his psychological reality as the source of infernal instincts quartered in the id, and therefore as the archantagonist of the superego. "Now what is the Devil, psychologically? The answer is eminently simple, on one level. *The Devil is the suspended superego.* . . . The Devil is that part of the person which permits him to violate the precepts of the superego."[78]

Although Professor Bakan's demonstration of Freud's antisocial contract is rather too academic to be representative of the conservative reaction (quite apart from the fact that its intention is not hostile but sympathetic to Freud), it serves to illuminate that suspicion of uncanny and subversive powers which leaves his rightist critics in no doubt as to where Freud's allegiance lay in the struggle between the demonic forces of the id and the spiritual resources of the superego. It is, instead, the "liberals" of Freudian criticism who are divided on the issue. On the one hand those who may be called the "left Freudians," broadly in sympathy with orthodox theory, interpret their text as a manifesto corroborating that of Marx against the repressive actions of society upon the freedom of natural (instinctual) man. In the vanguard of the left-Freudians are Herbert Marcuse and Norman O. Brown, who range themselves on the side of the instincts and look upon neurosis as "the essential consequence of civilization or culture." For these "instinctivist radicals," as Erich Fromm has called them, Freud is to be understood as an explosive social critic militantly in support of the revolutionary aspirations of the id toward freedom.[79]

On the other hand the *anti*-Freudians of the left, whose leading spokesman today is Fromm (followed by most of those erroneously designated as "neo-Freudians"), draw attention to other and different features of the master, both biographical and doctrinal, which present the portrait of an eminent Victorian secretly appalled by his own discovery of the carnal urges and forbidden wishes that flesh is heir to. In this anti-Freudian image the founder of psychoanalysis becomes clearly an exemplar, even a pillar, of nineteenth-century bourgeois mentality and morality— in effect, an ardent apologist for society and the superego.[80]

A corollary tendency of the left opposition has been to characterize Freud as the perpetrator of a quietist "Freudian ethic" counseling conformity and resignation to the mores of bourgeois-business society.[81] From this emphasis on social adjustment arises the familiar syndrome of the Organization Man, the other-directed character type who has escaped from freedom into the regressive security of the corporate womb. Support for this liberal indictment is found equally in the theory and therapy of Freud—and still more so in the practice of many orthodox psychoanalysts who evidently find it simpler to domesticate the patient than to underwrite a psychiatric "prescription for rebellion."[82]

Owing perhaps to the heat of the controversy between the two "left wings" of Freudian criticism, the possibility has not often been considered that *both* interpretations of Freud's position, how-ever logically contradictory, may be essentially correct—that in fact they may correspond to distinct but complementary aspects of his thought: i.e., those of *scientism* and *romanticism*. (From a perspective entirely external to the quarrel, that of the Catholic Church, this ambivalence appears more plausible; thus Jacques Maritain has observed of Freud's psychology that "it combines all the prejudices of deterministic, mechanistic scientism with all the prejudices of irrationalism. . . . But what can be said of a theory of the soul collaborated upon by Caliban become scientist and Monsieur Homais become irrationalist?")[83]

We have seen that Freud did indeed undertake, as earnestly as he knew how, to explicate human actions and aspirations in terms of determinate psychogenetic causes, and so in effect to deprive them of all autonomy and responsibility. Where his mode of explanation was most rigorously analytic and objective, Freud could give no quarter to illusions of creative striving or freedom of the will, let alone to the notion of a responsibly reasoning ego. On its mechanistic side his thought moved steadily in the direction of a corrosive pessimism with regard to the prospects of mankind generally and of their democratic aspirations in particular.[84] His science, in league with his inarticulate cultural bias, brought him down squarely on the side of civilization and against its discontented.

But we have seen that the writings of Freud also present, no less authentically, another and vastly different image—one that places him, as Reinhold Niebuhr among others has observed, "in perfect accord with Romanticism."[85] No one has more vividly evoked this shadow side of psychoanalysis than Thomas Mann in his essay of the 1920's, "Freud's Position in the History of Modern Thought":

> As a delver into the depths, a researcher in the psychology of instinct, Freud unquestionably belongs with those writers of the nineteenth century who, be it as historians, critics, philosophers, or archaeologians, stand opposed to rationalism, intellectualism, classicism—in a word, to the belief in mind held by the eighteenth and somewhat also by the nineteenth century; emphasizing instead the night side of nature and the soul as the actually life-conditioning and life-giving element; cherishing it, scientifically advancing it, representing in the most revolutionary sense the divinity of earth, the primacy of the unconscious, the pre-mental, the will, the passions, or, as Nietzsche says, the "feeling" above the "reason."[86]

If Mann's interpretation has any cogency, Freud was insistently aware of the significance of will and the passionate pull of purpose in the career of the person; his was an apocalyptic vision of the human struggle for existence as a matter literally of

Life and Death, of each against all, pitting the individual as tragic hero against the forces both of cultural necessity and biological fate. In the romanticist perspective, what had otherwise appeared as the conforming ethic of a coldly scientific determinism is transmuted into the moving poetry of classical drama.[87] It is not difficult to understand how this nocturnal visage should have come to engage the imagination not only of poets but of literary critics and classicists predisposed, it may be, by temperament and training to tragic intuitions of cultural crisis. What might be called the dramatistic appeal of Freudian psychology, as Mann's early essay suggests, has been there from the beginning; but it is noteworthy that it has had a much wider and deeper response in the years following the second world war than in the generation which came before. The earlier period, conditioned by the social conscience and reformist spirit of the thirties, was the heyday of revisionism in psychoanalysis—dominated by the hopeful figures of Fromm, Horney and Sullivan. In an age of reform the mind characteristically turns outward, away from somber introspection toward affirmative social action; evil then is likely to be perceived as a feature of lagging institutions, not as an irremediable flaw of the human heart. Freud's quasi-Hobbesian image of man's imperfectible nature was uncongenial with this progressive mood; and it soon gave way to the alternative image of Rousseau: man is born free, and everywhere he is in chains.[88] In the great task of social reconstruction, of breaking the chains, the overriding question addressed to psychoanalysts became simply: are you with us or against us? (The title of a *New Republic* article, "Must Psychology Aid Reaction?" well characterizes the Freudian criticism of this period.)[89] Either psychoanalysis was an active ally in the social struggle—or it was reactionary mysticism, giving aid and comfort to the enemy. In the generation of the New Deal, the custody and interpretation of the Freudian texts was by common consensus the prerogative of social scientists—whose ideas, in turn, were weapons.

Following the war and its chilling aftermath of cold-war disenchantment, the sanguinary mood of what Max Lerner has called "our lost Age of Confidence" was no longer tenable. Its place was taken by the thermonuclear era, the Age of Anxiety. The impulse to action became sicklied over with the recognition of complexity and ambiguity; and the cataleptic stance of brooding withdrawal once more came into fashion.[90] Social science abandoned the commitments of reform for the neutral value-freedom of behavioral science; academic philosophy abandoned pragmatism in order to hover between a positivist heaven and an "existentialist" hell; and psychoanalytic discussion, on at least its untechnical and public side, abandoned social psychology and cultural meliorism for what has turned out to be a full-scale romantic revival carried on predominantly (and appropriately) by students of literature.

The import of this belletristic reappraisal of the Freudian legacy, with its wholehearted resuscitation of the "night side" of psychoanalysis and its sweeping rejection of neo-Freudian rationalism, is illuminated in a polemical volume by a scholar of classics which has come to be almost the bible of the new criticism: Norman O. Brown's *Life Against Death*. The opening paragraphs of the book present a challenge and a manifesto:

> . . . It is a shattering experience for anyone seriously committed to the Western traditions of morality and rationality to take a steadfast, unflinching look at what Freud has to say. It is humiliating to be compelled to admit the grossly seamy side of so many grand ideals. It is criminal to violate the civilized taboos which have kept the seamy side concealed. To experience Freud is to partake a second time of the forbidden fruit; . . .
>
> . . . When our eyes are opened, and the fig leaf no longer conceals our nakedness, our present situation is experienced in its full concrete actuality as a tragic crisis. To anticipate the direction of this book, it begins to be apparent that mankind, in all its restless striving and progress, has no idea of what it really wants. Freud was right: our real desires are unconscious. It also begins to be apparent that mankind, unconscious of its real desires and therefore unable to obtain satisfaction, is hostile to life and ready to destroy itself. Freud was right in positing a death instinct, and the development of weapons of destruction makes our present dilemma plain: we either come to terms with our unconscious instincts and drives—with life and with death—or else we surely die.[91]

In the process of undertaking "a systematic statement, critique, and reinterpretation" of the body of Freudian theory, Professor Brown pauses to pay his disrespects to "the catastrophe of so-called neo-Freudianism"—which is, simply, the catastrophe of holding to the superannuated moral and rational beliefs of the liberal-democratic tradition. "It is easy," he writes in dismissal of the neo-Freudians,

> to take one's stand on the traditional notions of morality and rationality and then amputate Freud till he is reconciled with common sense—except that there is nothing of Freud left. Freud is paradox, or nothing. The hard thing is to follow Freud into the dark underworld which he explored, and stay there; and also to have the courage to let go of his hand when it becomes apparent that his pioneering map needs to be redrawn.[92]

This should be enough to indicate the direction of the romantic revival of Freudianism. It is notable not only for what it celebrates in Freud but also for what it relinquishes. In its denunciation of the "catastrophe" of neo-Freudianism it must also isolate and stamp out the germs of this nefarious line of thought (specifically any reliance on conscious reason) wherever they rear their heads in Freud himself. Since this indigestible element occurs most plainly and distressingly in Freudian therapy, the tendency of the new critics is to disentangle the therapy from the theory of psychoanalysis and then discard the former as irrelevant.[93] But so casual a severing of the intimate connections and reciprocal obligations which exist between Freudian theory and treatment is not only questionable from the standpoint of a rounded assessment of Freud; more immediately, it betrays the vast distance separating this essentially literary perspective from the clinical and professional center of psychoanalysis. As Maritain has pointedly observed, serious discussion of the problems raised by Freudianism has been immensely complicated by the fact that interest in the subject has spread far beyond psychological and psychiatric circles and "seems to grow greater and more ardent as it extends to less competent groups." Noting that "literary men have played an enormous role in the diffusion of Freudianism," Maritain asserts: "In the parasitical din that ensues it is seldom the voice of disinterested intelligence that is heard to advantage. All sorts of obscure desires of justification, vindication and a curiosity more or less pure intervene instead."[94]

In the current din of the romantic revival, whether or not it is the voice of disinterested intelligence which is being heard, other voices are in genuine danger of being drowned out. The sheer volume of the literature celebrating Freudian instinctivism and irrationalism obscures the fact that *it is moving in a direction diametrically opposite to the main current of psychoanalytic thought, both theoretical and therapeutic.* For to anyone conversant with the history of the psychoanalytic movement, not only "since Freud" but virtually since the first decade of the century, it is hard to overlook the fact that the trend has been strongly away from instinctual determinism (whether of the scientistic or romanticist version) toward rediscovery and affirmation of the creative, responsible, and consciously reasoning powers of the human mind.

Adler and the Great Departure

This trend is visible in nearly all the significant lines of thought charted by Freudian defectors, disciples and descendants alike over the past half-century. Despite their substantial differences in other respects, the major contributors to psychoanalysis (other than Freud himself) have with few exceptions displayed a common dissatisfaction with Freud's "mechanistic-materialist orientation,"[95] as well as with its clinical corollary of therapeutic neutralism. But this is to describe the departure from orthodoxy too narrowly and negatively; on its affirmative side, what is broadly shared by those who have had "the courage to let go of Freud's hand when it becomes apparent that his pioneering map needs to be redrawn" is the urge to restore the dimensions of wholeness and direction to the mental realm which Freud's analysis had effectively dissolved into blind and colliding fragments.[96]

This common point of departure, if nothing else, serves to

link such distinctive figures of the "first generation" as Adler, Jung, Stekel, Rank and Ferenczi. It is still more apparent, of course, among later revisionists such as Fromm, Horney, Sullivan, Schilder and Fromm-Reichmann; but it is also a prominent theme of many orthodox (or would-be-orthodox) Freudians of the thirties, most conspicuously in the important area of ego analysis. Finally, it is a widespread assumption of the most vital contemporary developments in *non*-Freudian psychiatry and psychotherapy, both in Europe (e.g., the existentialist school) and in America (the Rogerian school).

Alfred Adler's defection from the side of Freud, in 1910, may serve to mark the date of origin of this general departure in psychoanalytic theory and therapy. The specific disagreement between them grew out of Adler's resistance to the exclusively sexual derivation of the Freudian theory of neurosis;[97] but underlying this open dispute was a much deeper divergence of outlook and philosophy—at bottom, as Jung was perhaps the first to see, a difference in personal temperament. From the beginning Adler was a "holist," in the sense of the term which we have already examined in biology; and in fact the young Viennese physician may have brought this integrative orientation with him from his early studies in organic medicine.[98] As a psychoanalyst his professional concern (as Freud is said to have complained) was with the whole character of the patient as much as with his specific ailment.[99] From Adler's persistent effort to comprehend the individual in his unique ideographic "life-style"—i.e., on his own terms—came the title he was to choose for his own school: Individual Psychology.[100]

And from this personalistic emphasis also arose Adler's famous willingness to make use of various methods and approaches, however "unscientific" or unfashionable, which showed any promise of advancing therapeutic communication and cure—such as those of empathy, intuition and the phenomenological tradition generally. Finally, Adler's focus upon the individual in his ongoing *present* situation made him keenly aware of the significance of personal intentions and goals (as well as of biological drives and buried experiences) as crucial factors in the assessment of behavior.

"The most important question of the healthy and diseased mind," he declared, "is not whence? but whither? Only when we know the active, directive goal of a person may we undertake to understand his movements."[101]

Given his conviction that "the psychic life of man is determined by his goal," Adler was able, to a degree impossible for Freud, to *trust* the patient in therapy—that is, to give serious weight to his expressed opinions and conscious reasoning. This therapeutic confidence meant not only attending seriously to the patient's own insights and interpretations, but also respecting his capacity to understand and accept the truth. "Among my patients, children and adults," he could maintain, "I have never yet found one to whom it would not have been possible to explain his erroneous mechanism."[102] Adler, indeed, went a long way toward casting aside the protective mystique surrounding transference and countertransference, along with other interpretive devices calculated to detach the doctor from the patient; instead he sought to enlist the subject as an active partner in a mutual inquiry which has seemed to some critics rather more like a friendly dialogue than the conventional psychoanalytic examination. Psychotherapy, he maintained, "is an exercise in cooperation and a test of cooperation. We can succeed only if we are genuinely interested in the other. We must be able to see with his eyes and listen with his ears."[103]

Despite the charges of superficiality always hurled at him by Freudians, the influence of Adler upon psychoanalysis and psychology in general seems in retrospect to have been scarcely less extensive than that of Freud himself. The lonely course he embarked upon in 1910 has either anticipated or encouraged such vigorous developments as neo-Freudianism (or "neo-Adlerianism"), psychoanalytic ego psychology, client-centered therapy, existentialist psychology and contemporary personality theory.[104] From the perspective of our own day, it might even be argued (with conscious heresy) that it was the turn first taken by Adler some fifty years ago which has come to be the "mainstream" of the psychoanalytic movement—and that taken by Freud which has been in fact the "deviation."

The Dialectics of Becoming: C. G. Jung

The case of Carl Gustav Jung, the second of the great dissenters, is conventionally regarded as the antithesis of Adler's. Where, for example, Adler has been accused of preoccupation with the mental "surface" of the ego, Jung seems obsessed with a vision of the Unconscious as a bottomless (and unfathomable) abyss. Where Adler's straightforward language reflects the world of "common sense," Jung's exotic growth of imagery leaves him open to the charge of obscurantism and mystification.[105] Again, although both Freudian apostates share the distinction of excommunication, there is more evidence of hostility than of congeniality between their separate camps over the past five decades.

Nevertheless, these obvious differences are finally of less significance than the underlying theoretical assumptions and therapeutic aims held in common by Jung and Adler. Both staged their original rebellions against Freud's predominant emphasis upon a mechanical and overriding sexuality; both turned away from his rigid instinctual-infantilist determinism in favor of a teleology of future goals. Each of the defectors has been less concerned with scientific "analysis" of the patient's trouble than with his prospective *synthesis*, expressed in terms of the wholeness and integration of the total personality. Finally, for Jung as for Adler, the treatment relationship of doctor and patient is regarded as one of active and intimate transaction, a deeply personal encounter in which the commitment of the doctor is no less profound and consequential than that of the patient himself.

Perhaps because of Jung's numerous archaeological expeditions into the lower depths of the Unconscious (both personal and collective), it is often overlooked that he has consistently turned for guidance to the future no less than to the past, believing that "the mind lives by aims" as well as by causes. His emphasis, in both theory and therapy, is upon the creative potential of personality —the distinctive human capacity which he has been content to identify by the unfashionably archaic titles of "spirit" or "soul"—

as against the quest for scientific causality which in his view can only reinforce the "primitive tendencies" of the psyche. The one-sided method of analysis and reduction, Jung maintains, leads to a causal half-truth which can only induce resignation and hopelessness. "On the other hand, the recognition of the intrinsic value of a symbol leads to constructive truth and helps us to live. It induces hopefulness and furthers the possibility of future development."[106]

While Jung was far from denying the dynamism of instinctual themes, including the sexual, he undertook what amounts to a transvaluation of the values assigned to them by Freud—in particular, through his redefinition of libido as generalized psychic energy. "What I seek is to set bounds to the rampant terminology of sex which threatens to vitiate all discussion of the human psyche; I wish to put sexuality itself in its proper place."[107] Granting that the theme of sexuality is a morbid preoccupation of the modern world (much as when, suffering from a toothache, we can think of nothing else), Jung maintains that "it is an overemphasized sexuality piled up behind a dam," which shrinks to reasonable proportions once the way to normal development is opened. In fact, it is often the result of a blockage of more fundamental and productive energies of life striving for expression. "This being so, what is the use of paddling about in this flooded country?" Instead we should try to find the creative or constructive channels of expression which "the pent-up energy requires."[108]

Above all, according to Jung, we must avoid the temptation to partition and fragmentation of the human subject matter which scientific analysis encourages. The ultimate purpose of psychotherapy, after all, is not an objective analysis but a subjective synthesis; it is "the restoration of the total personality, . . . the bringing into reality of the whole human being—that is, individuation."[109] This internal process of individuation is one of purposeful and continuous development, the product of an indeterminate and life-affirming urge toward self-fulfillment—more precisely, toward the achievement of a fully mature and integrated self.

In his psychology as in his therapy, Jung has always insisted that *understanding* of the individual, as opposed to objective knowledge of his specifiable characteristics, is not to be reached by conventional scientific methods, which can only present "an abstract picture of man as an average unit from which all individual features have been removed."[110] Since it is just these individual features which are of paramount importance for understanding man, in his peculiar identity as a person, the first step toward that goal must be *to suspend the attitudes and preconceptions of objective science*.[111] Such a suspension of established belief, in the modern world, does not come easily; science has become virtually "the only intellectual and spiritual authority," and therefore "understanding the individual obliges me to commit *lèse majesté*, so to speak, to turn a blind eye to scientific knowledge." And if the psychologist happens also to be a doctor who "wants not only to classify his patient scientifically but to *understand him as a human being*, he is threatened with a conflict of duties between the two diametrically opposed and mutually exclusive attitudes of *knowledge*, on the one hand, and *understanding*, on the other."[112]

More even than Adler, Jung regards the therapeutic relationship as one of profound rapport, a "whole-souled" encounter in which the resources of both participants are fully engaged and which therefore threatens, as it challenges, the integrity of each alike.

> Two primary factors come together in the treatment—that is, two persons, neither of whom is a fixed and determinable magnitude. . . . The meeting of two personalities is like the contact of two chemical substances; if there is any reaction, both are transformed.[113]

While we may well expect the doctor to have an influence upon the patient, we need also to recognize the unfamiliar truth that this influence can only become real when he too is affected by the patient. "You can exert no influence if you are not susceptible to influence." It is not possible, even if it were desirable, for the therapist to detach himself from this personal involvement—"to shield himself from the influence of the patient and to surround himself with a smoke screen of fatherly and professional authority." The disturbing influence will be there in any case; the task is to bring it to consciousness and so transform it into a source of mutual enlightenment. Far from shielding himself, the analyst must become as "vulnerable" and susceptible as possible; in effect, *he must undergo his own treatment*. "The physician, then, is called upon himself to face that task which he wishes the patient to face."[114]

Jung's realization that the doctor "needs the treatment" no less than does his patient carries far-reaching consequences. Not only does it dissolve the "distance" between the two, but in a real sense abolishes their *difference* as well; the roles of "doctor" and "patient" can no longer support the conventional connotations of authority on one side and submission on the other. Therapy is for the "healthy" partner as well as for the disturbed; it is not merely a method of treatment but a means of self-understanding and self-realization: ". . . it is not only [for] the sufferer but the physician as well."[115] The demand of self-examination imposed upon the doctor by the very transaction of psychoanalysis gives it a relevance beyond the limits of the clinic; thus broadly redefined, the therapeutic relationship becomes a model of all interpersonal relations—of the human encounter everywhere in life. Nor is it the presence of scientific knowledge or medical expertise, for all their value, which gives to the process its distinctive "healing" quality: rather, it is the *understanding of the self and of the other* that arises from the therapeutic meeting of minds.[116] Indeed, Jung believes that the outcome of therapy may well depend more upon the characters of the two participants than upon "what the doctor says or thinks"; and he goes on to add still more pointedly: "The medical diploma is no longer the crucial thing, but *human quality* instead."[117]

The art and practice of psychotherapy, transcending the clinic, is therefore seen to furnish the basis for an orientation to life as a whole, to all the meetings of I and Thou—the foundation for a "normal" psychology of existence, addressed not just to neuroses of pathological origin but to the existential anxieties which make up the burden of freedom. "For as soon as psychotherapy requires the self-perfecting of the doctor," declares Jung, "it is freed from its clinical origins and ceases to be a mere method for treating the sick. It is now of service to the healthy as well, or at least to those who have a right to psychic health and whose illness is at most the suffering that tortures us all."[118]

As this remark suggests, the crucial difference between the psychoanalytic gospel according to Freud and the heretical revision of Jung lies in the latter's addition of the *prospective* factor—the transforming increment of choice and possibility, which might well be designated a psychic principle of indeterminacy. It is plain that, for all his idiosyncratic terminology, Jung accepts more or less (perhaps *more* rather than less) the whole dark body of subterranean forces and energies charted and codified by Freud; surely no one has been more insistent that there is an "animal" side to man's nature and a shadow side to his existence. But for Jung the scene is not wholly one of carnal struggle and the picture does not lie altogether in the shadows. The shadow, in fact, "belongs to the light"; the essential thing is "not the shadow, but the body which casts it."[119] For Jung there are two sides to man as there are two sides to everything in nature. Like Freud, he perceives the presence of dualities, the dialectical clash of opposites, wherever he looks in the world;[120] but, unlike Freud, his eye also takes in their harmonies and reconciliations. The dialectical presuppositions of Freud and Jung are, in this respect at least, comparable to those of Marx and Hegel. Freud, as we have seen, is very much the psychological Marxian, intent upon unmasking the antagonistic interests and reductively accounting for their actions in objective material terms. Jung in turn is closer to Hegel—not unmindful of the push of instinctual "matter" but confident also of the pull of "spirit."[121] For Jung, as for the author of The Phenomenology of Mind, the human spirit is not only a protagonist (thesis) in the psychic struggle but embodies the possibility of a higher synthesis of personal integration and self-fulfillment. It is on the basis of this dialectical idealism that Jung can unblushingly describe his own system as a "phenomenology of the spirit" and assert that what modern man seeks above all is not the release of his imprisoned sexuality, nor the omnipotence of scientific certainty, but simply his "soul"; i.e., the discovery of his authentic selfhood.

It is at about this point, in the ordinary encounter with Jungian literature, that the average reader throws up his hands. And admittedly it would seem to require an advanced esoteric curiosity to follow Jung and his disciples into the labyrinths of alchemy, Rosicrucianism and yoga—at the center of which reside what might be called the "Mendelian laws" of nature and life. But the essential insight, and contribution, of Jung's psychology is not deeply dependent upon the findings of these anthropological and theological excursions. That insight may perhaps be conveyed in the image of man as an indeterminate composition of lights and shadows, an unfinished sculpture with the singular capacity of choosing and carving out its own final form—in effect, whether to rise above the clay or to harden permanently within it. The contending opposites in the personality which Jung has defined in the old-fashioned terms of "nature" and "spirit" would seem to bear a close resemblance to those distinct but complementary dimensions of human behavior which we have identified earlier as the *organic* and the *humane*. For the "spirit" that Jung has in mind implies no divine existence or supernatural faith but only the well-accepted concept of the *human spirit*, the spirit of humanity, whose aspirations have historically found expression in works of reverence and revelation. And, on the opposite side, the "nature" which he has in view is none other than the familiar physical nature presented by organic evolution and biological endowment—the measurable side of man which contains all his necessary mechanical equipment of glandular substances, palpable impulses and quantitative energies. The opposition between these two modes of existence—organic nature and the human spirit—is clear to Jung, and with it the urgent fact of *choice*. The first mode points to "the inexorable cycle of biological events"; the other opens a path to freedom. "There is nothing that can free us from this bond except that opposite urge of life, the spirit. . . . It is the only way in which we can break the spell that binds us to the cycle of biological events."[122]

Given this assertion of a psychic principle of indeterminacy, it is not surprising to find alongside it what one disciple has called Jung's "fundamental principle, the law of *inevitable complementariness*, according to which psychological happenings must occur."[123] The psychology of Jung, in theory and therapy, documents and illuminates more impressively than any we have yet observed that evocative analogy advanced on other grounds by Bohr and Heisenberg: namely, that in the study of his own life and kind, of himself and the other, man is at once an actor and a spectator—a participant-observer—who cannot, in the interest of truth, remain a stranger to himself. To the science of *causes* (the "inexorable cycle of biological events") constructed by Freud, Jung has sought to add a science of *aims*; to the therapy of psycho-analysis he has added a program of psycho-synthesis; and against the assumption of genetic determinism he has contributed an avowal of self-determination. For Jung, as Ira Progoff has seen, "man is caught in a trap only if he permits himself to be."[124] Jung's hypothesis of Individuation represents the dialectics of becoming—a system of humane evolution which holds out to modern man a genuine choice of possibilities on the basis of a faith in the possibility of choice.

The Existential Will of Otto Rank

The essential point of this review of the contributions of Adler and Jung might be simply described (in a phrase already familiar) as *respect for the person:* respect in terms of theory for his powers of creativity and responsibility, respect in terms of therapy for his values, his intentions and (above all) his peculiar identity. It might even be suggested, in the language of the clinic, that the growth of the psychoanalytic movement has been deeply influenced by a traumatic experience suffered early in its career: namely, the shock of recognition induced by the discovery of the human being as a subject (Thou) rather than an object (it)—as a singular personality, ultimately indeterminate and impervious to the probe of objective science.

This recognition of *man-in-person,* as opposed to *man-in-general,* carries with it an awareness of the insufficiency of the classical image of science as enshrined in the canons of detached observation and reductive analysis. In the burgeoning field of psy-

choanalysis, as in its companion realms of perception and psychobiology, numbers of responsible students have found themselves
drawn to the unsettling conclusion that the most distinctive and
definitive features of a human being cannot be made out at all
from a "psychological distance," but can be brought into meaningful focus only by an understanding of (literally, by *standing
under*) the unique perspective of the individual himself.

Few psychoanalysts have been more emphatic on this score
than the third of those dissident members of the original Vienna
circle who felt themselves compelled to break with the faith. Otto
Rank, the youngest of the three and the last to take his departure,
carried even further than his predecessors the heretical themes of
individuation and self-determination, of the creative potentialities
of mind and will, and of the intrinsic uncertainty and spontaneity
of the therapeutic adventure.

One of the paradoxes of Rank is that, although he began
his career as the most theory-minded and "metaphysical" of the
circle—and was evidently regarded by Freud as its unofficial philosopher[125]—he was drawn more and more to concentrate his attention upon the highly personal dynamics of the therapist-patient
relationship, and in the process to cultivate a sharp distrust of general principles and theoretical mandates.[126] In turn, however, in the
very labor of giving expression to his intuitive understanding of the
treatment relationship, Rank found himself developing an independent theory of human nature—an original and affirmative image
of man as the maker, and executor, of his own will.

The mutual obligations of Rank's therapeutic method and
his psychological theory are suggested by the fact that his two
most systematic works, one on therapy (*Will Therapy*) and the
other on theory (*Truth and Reality*), were composed simultaneously. (Another indication is the designation of his method of
treatment as a "Philosophy of Helping.")[127] What is more significant is that, while Freud in the 1920's was coming increasingly to
subordinate his own clinical interests to the higher calling of
theory construction, Rank was moving in exactly the opposite
direction—to such an extent that his psychological speculations

came to seem the barely varnished projections of his experience in the consulting-room.

An illustration of this may be seen in the basic Rankian concept of "relationship." In its ultimate theoretical form the term came to signify all those influences upon personality development which were later familiarized by Sullivan under the rubric of interpersonal relations—more particularly, the relationship with a "Significant Other." Thus Rank maintained that "the ego needs the Thou in order to become a Self. . . . The psychology of the Self is to be found in the Other, . . ."[128] The original source of this concept was the helping relationship of therapist and patient. In that curious sustained intimacy, Rank perceived both the limitations of reductive scientific methods and the clue to a deeper understanding between man and man. In short, the term "relationship," at first a reference to a special technique of treatment, became in the end a general hypothesis about human nature and a prescription for its self-activated development.

Rank attributed his own departure from orthodoxy largely to dissatisfaction with Freud's slighting of the more personal therapeutic side of the movement in favor of the remoter interests of scientific theory. In particular—and with a vehemence unmatched by either Adler or Jung—he assailed Freud's theoretical reduction of human behavior to involuntary causes rooted in biology and infancy. At times Rank became so exercised about this scientism that he could seem to be condemning "science" itself as an entirely irrelevant and pernicious influence. But for the most part he made clear that his objection was to the wholesale transfer of the methodology of physical science to the subject matter of psychology— "the characteristic of which is just that it can't be measured and checked and controlled."[129] What seemed an assault against science per se was in reality an attack upon a one-sided image of science, launched from the point of view of an alternative scientific ideal: "In interpreting the human element scientifically, psychoanalysis had to deny it and so defeated its own scientific ideal, becoming unscientific by denying the most essential aspect of the personality."[130]

Far from opposing science in principle, Rank undertook to bring his own perspective into line with the changing orientation of natural science as expressed in relativity and quantum theory. Describing his approach as a "relativity theory" of psychology, he saw a parallel in the dissatisfaction of the new physics with the traditional bifurcation of subject and object—of consciousness on the one hand and an objectively independent reality on the other. The contemporary physicist, Rank maintained, sees nature very much as does the modern (or postmodern) psychologist; both have had to free themselves from the framework of a "natural science ideology" which ignored the instrumental role of consciousness.

> On the basis of the same ideology which became so fateful for physics, Freud had fallen into the error of trying to discover behind mental phenomena a reality free of consciousness, and of believing that he had found it in the unconscious. He regarded his interpretation of the world as reality itself. But with this attempt to comprehend mental phenomena objectively, it became completely obvious that such comprehension could be mediated only by consciousness, irrespective of the subjectivity or objectivity of the observations made.[131]

There is more to Rank's "relativity theory" than this insistence upon the intrusive fact of consciousness. Just as it may be argued that there is no really objective history (since each generation views the past through the window of its own present), so Rank considered that each age also carves out its appropriate "psychology" (or philosophical anthropology): in effect, it recreates man in its own image. Freud, with his causal explanations and sturdy posture of detachment, had hoped to construct an objective science of psychology, free from any taint of moral or religious ideology. But the paradox, Rank saw, is that Freud's scientific psychology had itself become an ideology—a substitute for traditional religion and morality.[132] Far from transcending its cultural origins, it came to reinforce the ideals of adjustment and conformity which characterize the bourgeois outlook. Like modern educational theory, Freudian psychoanalysis has been turned into an instrument of social control, molding the individual into a peaceful and acquiescent citizen, without a will of his own. Thus the psychological "revolution" which Freud thought himself to be inaugurating has become, in its social consequences, a counterrevolution.[133]

What Rank was arguing was that the Freudian image of "Psychological Man" is no less partial and culture-bound than the anthropological models it sought to supersede—whether Darwin's Biological Man, Marx's Materialist Man, Adam Smith's Economic Man—or the Mechanical Man of the left Cartesians. Each of these antecedent images had made a scientific claim to objectivity and completeness; each now stood revealed as a historical curiosity, an obsolete museum piece. For Rank the lesson was plain: every attempt to comprehend human vitalities in terms of mechanical causes and objective conditions is self-defeating; all that it can ever illuminate is the ideological bias and partial blindness of its author.

The inference was not merely that a "natural science" of psychology is impossible. Rank had a more positive point to make. If psychology could not overcome its historical relativity or shake off its ideological trappings, then it should realistically confront its limitations and carry out its assigned mission. It should become in earnest a "mirror for man," reflecting a world view genuinely consistent with the highest capacities and purposes of mankind in its own stage of cultural development. Rather than seek an impossibly objective and universal system, psychology should recognize the need for periodic changes in theory to meet altered conditions and new necessities.[134]

Moreover, there is a deeper incentive for man to proceed to "make" his own psychology: namely, because he has the existential capacity to *make himself*—in Rank's phrase, to "will himself"—and so determine his own fate. This is what Rank's ambiguous will-psychology affirms at bottom: the human right of self-determination and self-realization, the possibility of conscious becoming. Men are born free; if nevertheless they are psychically in chains, at least they possess the power to be born again. The symbolism of birth and rebirth always held a compelling fascination for Rank; his early theory of the trauma of birth—which drew down the wrath of Freud but has since gained in credibility—came to acquire the broader meaning of a metaphor expressing the successive crises of *choice* in human experience, "the never completed birth of individuality." For Rank "the whole consequence of evolution from blind impulse through conscious will to self-conscious knowledge, seems still somehow to correspond to a continued result of births, rebirths and new births. . . ."[135]

To be sure, not many men are prepared to face the challenge of themselves, to assume the full responsibility for their own existence. Rank concluded that there were three levels or styles of response to this self-challenge: the first, and most common, was simply to evade it; the second was to make the effort of self-encounter, only to fall back in confusion and defeat; the third, and much the least common, was that of carrying the confrontation through to self-acceptance and "new birth." These three attitudes or approaches correspond to Rank's three types of human character: the "average" or adapted man, content to swim adjustively and irresponsibly with the tide; the "neurotic" type, discontented alike with civilization and himself; and the "creative," the twice-born (as represented in the ideal-types of Artist and Hero)—at peace with himself and at one with others.[136]

On closer examination, these three character types seem reducible to two: the adjustive and the creative. The intermediate "neurotic" type is more accurately described as a stage of arrested creative development—what Rank called the *artist-manqué*. Thus Rankian theory presents two alternative approaches to the self and the world (of culture): that of unconscious rejection of the self and acceptance of the culture, or that of conscious acceptance of the self and rejection of the conventional culture. The first route is the familiar one of escape from freedom, the straight and narrow path of least resistance. The second is the way of creative growth and becoming—the road to autonomy. The task of therapy for Rank is primarily to make clear these alternatives and to help in the liberation of the patient's own will for the voluntary act of choice: to be (himself) or not to be.

But the therapeutic encounter is also more than that: it is a *relationship* whose aim, paradoxically, is *separation*. Rank fully understood the vital connection between independence and *inter*-dependence—the fact that the human person is brought to maturity not through alienation but through affiliation. "The ego needs the other," at each successive stage of self-development, in order to be confirmed in the recognition of its own identity.

There is in all of this a close resemblance to the Hegelian dialectic of assertion, opposition and union.[137] The individual will (in childhood, in adult life, or in therapy) needs what Rank termed a "counterwill" against which to declare itself and find expression: an antithesis which it discovers typically in the parent, the lover, and the therapist. The synthesis resulting from this opposition of wills represents a new and higher integration, accomplished through an experience of relationship which an older vocabulary knew as *love*. "The mature person," as Ruth Munroe has written in a summary of Rank's view, "loves himself in the other, and the other in himself. Awareness of difference, of partialization, enriches the new sense of union. This union is not the effortless bliss of the womb, but a constantly renewed creation."[138]

A central feature of Rank's "helping" therapy was his insistence upon subordinating the will (and the interpreting mind) of the therapist to that of the patient. The essential task became that of *freeing* the patient's will from the coercions alike of instinct (id) and of culture (superego). Rank's system was, in contrast to Freud's, virtually an "ego-psychology," a psychology of consciousness. Through all the shifts and transitions of his developing perspective—even when, in his last years, he searched beneath consciousness for deeper sources of willing—Rank's most persistent theme remained this stress upon the liberation of the human powers of creativity, responsibility, autonomy. Thus he wrote in his unfinished last work, *Beyond Psychology*, that it was on this "vital issue" that he had deviated from Freud's mechanistic conception of the ego as the puppet of forces beyond its control.

The whole question of psychological therapy resolves itself, in the last analysis, to the philosophical problem of a deterministic versus a vitalistic point of view. But even if psychology could be conceived of scientifically in terms of strict determinism, psychotherapy as a living process of personality development can never be based on a deterministic point of view. In trying to establish what the individual is, and not what has happened to him, constructive therapy does not aim at adjustment but strives to develop autonomy in the individual, thereby liberating his creativity.[139]

Psychosynthesis: The Post-Freudian Consensus

The title that Freud bestowed upon his science of the mind was remarkably appropriate. Psycho-analysis, in its classical formulation, meant just that: the reductive *analysis* of the psyche, conducted from a scientifically detached posture. On the other hand, the central characteristic of the great departure from Freudian psychoanalysis may best be indicated by giving it a distinctive title of its own: that of *psychosynthesis*.[140] For the persistent effort of this countermovement over the years has been to join together what Freud had systematically put asunder—to restore unity and wholeness to the partitioned psychic realm, and to do so not from a therapeutic distance but from the close quarters of personal encounter and affiliation.

A basic premise of the classical school stands out starkly from the title of one of its representative works: Karl Menninger's *Man Against Himself*. The contrary assumption of the "neo-Freudian" school is conveyed by one of its own most celebrated works: Erich Fromm's *Man for Himself*. But even this contrast of themes is incomplete; for the Freudians man is not only (by nature) against himself, but against all others. For the neo-Freudians, man is not only (by nature) *for* himself, but *with* all others. In the post-Freudian consensus, man's deepest need and possibility lies in the mature form of relationship or "relatedness" to others that Fromm calls productive love—which carries with it, in turn, the birth of individuality and the realization of the self.

It is already apparent, from our survey of the great dissenters, that this synthetic or holistic approach to personality was under way long before the advent of the revisionist school of the thirties. In addition to the triumvirate with whom we have dealt, two other members of the original psychoanalytic circle deserve to be mentioned whose sins were of a similar order—and whose ex-communications were, if anything, still more dramatic.

Wilhelm Stekel and Sandor Ferenczi were both in the course of time to be judged guilty of the same professional "crime": that of undue sympathy with the patient. More precisely, in place of the austere passivity of Freud's therapeutic posture, Stekel and Ferenczi substituted an approach of active involvement and con-cern—in the apt phrase of Izette de Forrest, "the leaven of love."[141] In Stekel's case (the earlier of the two), this engagement carried with it other unorthodox insights: among them a recognition of the therapeutic relevance of the analyst's own personality and values; an emphasis on the "prospective factor" or future aspirations of the patient as an aid to understanding; and finally an appreciation of his ongoing conflict in the "here and now" as opposed to its sources in the distant past.[142]

Ferenczi carried this sense of commitment and responsi-bility a step farther. Not only is the therapist engaged "actively" in the treatment, but also personally and authentically. Not only is the patient's conflict a reality of the present, but so is his relation-ship with the therapist. In this view it is not two "roles" or shadow-players that meet in the analytic hour, but two human beings— more accurately, *these two* human beings. As a disciple of Ferenczi has written:

> The truth is that in the analytic consulting room there are two people, each living vital lives, each bent on solving one and the same problem, meeting day after day for several years, growing to know each other better with every day. It is impossible to imagine and ludicrous to assert that an emotional relationship on both sides must not inevitably develop in such a setting. It is outside the realm of possibility that an analyst who is sincerely determined to cure his patient does not grow to care for him.[143]

This voluntary abandonment of the "blank screen" of an-alytic objectivity seemed to Freud not only unscientific but virtu-ally indecent.[144] It may be that, in their enthusiasm for "active" therapy, Stekel and Ferenczi tended to become overactive; and no doubt they overstepped the conventional boundaries of professional decorum. But their willingness to put aside the protective masks and armor of their profession, to close the gap between analyst and patient—in short, to become actors as well as spectators—established the precedent for a succession of related developments in psychotherapy, ranging from the strenuous activism of Wilhelm Reich through the psychodramas of Moreno to the participant-observation of Sullivan and his school.[145] The degree of sympathetic engagement first taught and practiced by Stekel and Ferenczi no longer strikes the informed student of therapy as remarkable; on the contrary, in the light of subsequent trends, their affective and affectionate turn toward the "other" appears today as comparatively modest and almost diffident.

The therapeutic innovations of these two early members of the movement, like those of the more famous deviationists discussed above, stand at a midpoint in time between the Freudian orthodoxy and the neo-Freudian heterodoxy. Before turning to the fundamental changes rung upon psychoanalytic theory and therapy by the revisionist generation, it is in order to take note of an independent line of thought which contributed substantially to the "great departure" from classical doctrine but which somehow managed to be separate without being regarded as separatist: i.e., the psychoanalytic school of ego psychology.

In a notable article of the 1940's, "The Ego in Contemporary Psychology,"[146] Gordon W. Allport called our attention to one of the strangest events in the history of modern psychology: "the manner in which the ego—or self—became sidetracked and lost to view." The one psychological fact about which, it might be thought, everyone could feel confident—the reality of his individual self or ego—had come to be systematically overlooked, rejected or disparaged. The prime mover in this anti-egoistic trend was, of course, behaviorism and its positivist variants. But, as Allport pointed out, the movement also drew support from a rather surprising source: Freudian psychoanalysis. For despite its great attention to the concept, psychoanalytic theory deliberately minimized the scope of the ego and held its authority in low esteem. Partly conscious and partly sunk in unconsciousness, the Freudian ego was a thing of elaborate defenses but no dynamic power, seeking more or less vainly to mediate the conflicting demands of the id, the superego and the outside world—"essentially nothing more than a passive victim-spectator of the drama of conflict."[147]

Freud himself came to modify this harsh appraisal late in his career, but never to the extent of granting to the ego genuine autonomy or any capacity for self-initiated activity toward the world. "It was hard for him to be an ego theorist," as an unfriendly critic has remarked, "when at heart he was an ego iconoclast."[148] More substantial and constructive inroads were made by Adler, for whom the conscious ego became a paramount consideration in theory and therapy alike; and later by Rank in the development of his theme of conscious self-actualization. With Rank the "ego ideal" was separated from its dependence on the superego—at least for the creative personality, who "evolves his ego ideal from himself, not merely on the ground of given but also of self-chosen factors which he strives after consciously."[149]

Within Freudian orthodoxy the first major step beyond the classical formulation was taken by Anna Freud, in *The Ego and the Mechanisms of Defense* (1936). Where formerly the emphasis had been upon the more clearly defined materials of id and superego, Anna Freud demonstrated that the ego also had deep-lying roots and important consequences of its own for personality development.[150] As the title of her book indicates, however, the role and character of the ego remained essentially defensive (if not "mechanistic"), subordinated alike to the superior authority of the superego and the infernal power of the id.

A longer stride forward was taken by Heinz Hartmann, since regarded as the father of the modern ego school, beginning with his paper "Ego Psychology and the Problem of Adaptation" (1939).[151] Although he sought to operate within the accepted matrix of Freudian libido-instinct theory, Hartmann gave to the ego and its "apparatuses" a twofold autonomy from the other psychic agencies of id and superego. First of all, there was "primary autonomy": the ego was to be seen as an original datum of human evolution and personal character, rather than merely an outgrowth of the id; it was thus essentially independent and coequal in status.[152] Hartmann posited a "conflict-free" sphere of ego activity operating from birth on, which lay outside the range of conflicts and drives stemming from the id. But he also added another dimension of "freedom": even where the ego does become bound up with instinctual-erotic drives, it acquires in the course of personal growth a "secondary autonomy"—that is, its activities take on meanings and functions independent of their original motivation. These "neutralized" patterns of behavior are therefore, in their maturity, no longer explainable in terms of the initial reason-for-being; they have left their causes behind, as it were, and struck out on a career of their own.

This is indeed a radical shift of emphasis—although it has not always been regarded as such. It brings the ego psychology of psychoanalysis directly to the theoretical point which was reached at almost the same moment by Gordon Allport—a point which furnished the departure for the construction of Allport's affirmative and "personalistic" theory of personality. It is striking to note the close similarity, in language as well as concept, between Allport's theory of the *functional autonomy of motives* and Hartmann's theory of *secondary autonomy*. And, noting this, it is difficult to avoid the conclusion that ego psychology—hardly less than Allport's psychology of becoming—paves the way for a positive appreciation of the human ego (with its apparatuses of perception, thought, imagination and purpose) as a psychic agency intrinsically responsible, reasonable and respectable.[153]

The Freudian view of motivation, which reduces every striving to a moment in its source, may indeed, as Allport has pointed out, furnish an acceptable model for *neurotic* behavior while being wholly inappropriate for *normal* behavior. It is almost a truism to say that the most obvious characteristic of neurotic behavior is its compulsive or "driven" quality. Such behavior is predictable precisely because it is *determined*—because it is patterned or stereotyped, marked by repetition, fixation and rigidity. But to recognize these as the distinguishing symptoms of *abnormal* behavior implies, both logically and psychologically, an altogether distinct model for normal behavior. Such a model is that presented by Allport (and now broadly shared with a small army of colleagues). His alternative view, that of the functional autonomy of motives, holds that "motivation may be—and in healthy people usually is—autonomous of its origins. Its function is to animate and steer a life toward goals that are in keeping with *present* structure, *present* aspirations and *present* conditions."[154] Something of the difference which such a theory of ego-autonomy entails for psychology, and for the explanation of behavior generally, has been suggested by Allport in a summary statement:

> It is necessary only to insist that the forward thrusts of motivation that are so characteristic of human personality cannot adequately be accounted for by any doctrine of pushes, even a sequence of pushes, out of the past. An adequate theory must allow for the effectiveness of a current self-image and for the dynamic character of intentions, of value-orientations and of uniquely patterned psychogenic interest systems in normally healthy adults.[155]

As this indicates, there has been a definite convergence of outlook between the ego psychology nourished under psychoanalytic auspices and the personality psychology of such contemporaries as Allport.[156] Both schools have reacted against the tendency to ignore the *minding person,* and more particularly the conscious ego, in rapt concentration upon his specifiable machinery of drives and reflexes. In the classical system of psychoanalysis, there has never been much serious use of (or use for) the concept of the *self*—that familiar construct of nineteenth-century psychology which subsequently fell under the same obloquy as the more venerable notion of "soul."[157] But in the last generation the self has made a remarkable comeback—under new management.

Quite possibly the most significant of all the "revisions" performed upon the body of Freudian theory by the so-called neo-Freudian school has been its rediscovery and rehabilitation of this old-fashioned idea of the self. Although the formulations vary, the term is equally central to the vocabularies of Erich Fromm, Karen Horney and Harry Stack Sullivan—by general consensus the leading figures of the school. In particular, Fromm's concept of the "true self" and Horney's equivalent notion of the "real self" convey this shared recognition; and so, in its own way, does Sullivan's specialized construction of the "self-system" (although, to be sure, as in the earlier psychologies of Mead and Dewey, the figure of the self in Sullivan's "Gestalt" is sometimes hard to distinguish from its ground).[158]

These revisionists also share another insight with regard to the self which separates them further from Freud—that which was anticipated by Rank in his observation that "the self needs the other." This recognition, of immediate relevance to therapy as well as to theory, finds expression among the revisionists in an insistent awareness of the importance of *relationship* in the growth of personality. In the broadest sense, this emphasis implies the general and reciprocal interaction of "culture and personality"—an acknowledgment which gained great impetus in all of social science during the thirties, owing to such influences as the new cultural anthropology and the deepened "social consciousness" of intellectuals in the depression period.

More intimately and pertinently, the stress on relationship has conveyed an appreciation of the role of personal affiliations and attachments in the career of the individual—his basic need for "relatedness" with others (Fromm), and especially with "significant others" (Sullivan). For Sullivan, influenced by Sapir and Mead, psychiatry itself was basically to be understood as the "science of interpersonal relations"; for Fromm, trained in European social psychology, psychoanalysis became ultimately the science, and art, of "*loving*."[159] Horney, no less than the others, discovered her key metaphors in the sphere of personal contacts; through most of her career she regarded neurosis as fundamentally a "disturbance of human relationships," eventually modifying this somewhat to incorporate greater emphasis upon factors within the self as well as outside it.[160]

Like Adler, Horney was emphatically a *holist*: her concern

as a theorist was less analytic than synthetic, as expressed in the favorite phrase "the whole patient in therapy." With this sense of personal wholeness went a corollary accent on individual *uniqueness*.[161] Consistent with this conviction, Horney's animus was directed principally against the mechanistic analysis of the patient attributed to Freudian doctrine. Such objective dissection, in her view, could never flower into recognition of the peculiar identity of the human being under treatment; hence it could never bring the therapist to the sympathetic understanding of the patient through which he might in turn be led to "self-analysis" and self-realization.[162]

The avowedly cultural and interpersonal orientation of the neo-Freudians is, no doubt, sufficiently well-known and understood to need little demonstration. Much the same might be said (Sullivan possibly excepted) for their equally conspicuous reliance upon the creative, responsible and potentially autonomous capacities of man.[163] Where such older rebels as Jung and Rank have been grossly misunderstood (or simply unread) on this side of their theories, and therefore stand in need of explication, there is little dispute regarding the *presence* of these themes in the writings of the main neo-Freudians. On the contrary, as noted earlier, the criticism most often leveled against them is that they are much *too* "rationalistic" and "optimistic": either that they are psychological innocents, denying evil and rationalizing sin—or (less innocently) that they are latter-day Samuel Smileses dispensing patent nostrums of self-healing and self-help. It is this evaluation which leads Norman O. Brown, for example, to pronounce sweeping judgment upon the neo-Freudian movement as a "catastrophe."[164]

If it is not difficult to perceive how this bitter charge has arisen (in part from the almost necrophilic romanticism of the new critics, in part perhaps from the deceptive simplicity of Fromm's literary style), it is more difficult to know how the objection can be seriously sustained. No one in the entire psychoanalytic tradition, including Freud himself, has been more insistently concerned with the modern dilemmas of alienation and anxiety than has Horney—unless it is Fromm. The escape from freedom, the nightmare of neurotic anguish, the self-contempt and self-hate of the driven or dominated (sado-masochistic) personality, the compulsive submission to irrational authority and the automaton urge to conformity: these are the familiar and recurrent, if not quite dominant, themes throughout the writings of both Horney and Fromm.[165] If they are not finally dominant (as the neo-orthodox romantics would have them be) it is because the dilemmas they describe are not regarded as inevitably fatal: i.e., they do not exhaust the possibilities and resources of human experience. (In this respect, Sullivan is a separate case mainly in that his peculiar vocabulary is drawn from other traditions, such as those of medical science and American pragmatism; but an adept at "Sullivanese" has no difficulty finding conceptual equivalents for each of the themes recited above.)[166]

The post-Freudians, in short, are far from guaranteeing a happy ending to the human "situation tragedy." What they offer is rather the hard contingency of an open psychological system—the "dreadful freedom" of an indeterminate existence, which carries threat no less than it carries opportunity. In this perspective the achievement of freedom—the act of liberation—is always an estrangement, both biographically and historically. From the traumatic emancipation of birth onward—or from the traumatic emergence of the individual out of the tribe—freedom is encountered in fear and trembling. The wish to escape from this separation-anxiety takes a variety of forms. Psychoanalytically it may be expressed in terms of regressive behavior; the womb is totalitarian but certain in its security. Politically it may be expressed in authoritarian behavior; the total state is womb-like but secure in its certainty. For personalities "intolerant of ambiguity" —unwilling to confront the existential risks of chance and choice —even the certainty of death may be preferable to the uncertainty of existence (better a dead coward than a live hero). Herein, it may be, lies much of the appeal, and the plausibility, of Freud's Thanatos: the death-instinct conveys the wish for a surcease of struggle in the peace of the grave—the organic urge to return to the inorganic.[167]

But the organic urge (even if it be granted) is not the human will. Where Freud was committed by his premises to an explanation of behavior in determinate *organic* terms, the neo-Freudians present a complementary viewpoint framed in indeterminate *humane* terms. (Here, as elsewhere in science, the recognition of complementarity is implicit in the acceptance of uncertainty.) There can be no question that the Freudian vision is more deeply pessimistic and thoroughly fatalistic than the neo-Freudian re-vision; but it is at least debatable which has the superior claim to embodying a genuinely *tragic* vision. The difference between the two, on this level of eschatology, has nowhere been better defined than in the concluding sentences of the work which turned out to be Karen Horney's last testament. Pointing out that it is the characteristic of man that his reach should exceed his grasp, she observed:

> This in itself is not a tragic situation. But the inner psychic process which is the neurotic equivalent to healthy, human striving *is* tragic. Man under the pressure of inner distress reaches out for the ultimate and the infinite which—though his limits are not fixed—it is not given to him to reach; and in this very process he destroys himself. . . .
>
> Freud had a pessimistic outlook on human nature, and, on the grounds of his premises, was bound to have it. As he saw it, man is doomed to dissatisfaction whichever way he turns. He cannot live out satisfactorily his primitive instinctual drives without wrecking himself and civilization. He cannot be happy alone or with others. He has but the alternative of suffering himself or making others suffer. . . .
>
> Freud was pessimistic but he did not see the human tragedy in neurosis. We see tragic waste in human experience only if there are constructive, creative strivings and these are wrecked by obstructive or destructive forces. . . . In most general terms, what we regard as a healthy striving toward self-realization for Freud was—and could be—only an expression of narcissistic libido.
>
> Albert Schweitzer uses the terms "optimistic" and "pessimistic" in the sense of "world and life affirmation" and "world and life negation." Freud's philosophy, in this deep sense, is a pessimistic one. Ours, with all its cognizance of the tragic element in neurosis, is an optimistic one.[168]

Psychotherapy: Toward the Meeting of Minds

Progress in psychotherapy will depend to a large measure upon whether or not we can free ourselves from the dictates of science. The imposition of rules which fit the outside observer has to be rejected, and inner experience must be restored to the position it has held for thousands of years. The outer and the inner observer stand in a relationship of complementarity. What one sees, the other does not see, and vice versa. This acceptance of [the] limitations of the human observer brings us into agreement with the views of modern physicists.

—Jurgen Ruesch[169]

If the neo-Freudians were the agents of revision on the theoretical side of psychoanalysis, they were the catalysts of a major rebellion on its practicing side. The broad outlines of this radical change in perspective, centering upon the relationship of analyst and patient, have been pre-figured in our discussion of the early dissenters. Their preliminary and isolated revolts—directed, inter alia, against the reflecting-mirror or blank-screen image of the analyst, against the arbitrary convention of the couch, against the elongated time-span of analysis "terminable and interminable," and against the dogmas surrounding transference and countertransference—assumed in the thirties the distinct proportions of an organized and articulate movement. After the insurgent movement had run its course, little remained as it had been in the consulting-rooms of the house that Freud built—or on the premises surrounding it.

To be sure, it might still be argued that the important changes in the technique of therapy were "evolutionary" rather than revolutionary (i.e., consistent with the development of classical conceptions), and even that all of them had somewhere been anticipated in the voluminous pages of the Collected Papers. But these are fine points of doctrinal sensitivity which need not detain us. Our concern is only to demonstrate that certain fundamental and far-reaching changes were gradually brought about in the therapy of psychoanalysis, corresponding to deep inroads in its theory, which have drastically altered the images of patient and doctor in turn, as well as of their peculiar relationship.

More specifically, what requires to be shown is that the technique of psychotherapy nurtured within this tradition, in the process of its growth into ever more systematic and accepted practice, has evolved in the direction not of greater objectivity and causal analysis, but rather of "intersubjective" encounter and recognition of the patient as a *Thou*—on the intimate plane of his present existence as well as of his "life plan" for the future.

On this score the evidence, if not unanimous, is undeniably compelling. Enough has perhaps been said about the neo-Freudians to indicate their general contribution. Its crucial aspects have been aptly characterized by Fromm in his definition of the "syndrome of attitudes" making up what he calls the productive orientation, a syndrome which finds expression alike in personal behavior and in the therapeutic situation: namely, "*care, responsibility, respect and knowledge.*"[170] For the therapist each of these subjective qualities (including even knowledge in its ultimate sense of understanding) manifestly enforces an obligation of active participation and self-commitment—not to say of "love."

It was Harry Stack Sullivan, among the neo-Freudians, who gave the greatest impetus to the therapeutic value of participation through his conception of the psychiatrist as a *participant-observer*. "The therapist has an inescapable involvement in all that goes on in the interview," he insisted; "and to the extent that he is unconscious or unwitting of his participation in the interview, to that extent he does not know what is happening."[171] No less than Fromm, Sullivan heavily underlined the therapeutic attitudes of *care* and *respect*; and it is especially striking that his warm appreciation for the personal worth and responsible capacity of the patient grew out of his work with psychotic individuals—those most "irresponsible" of persons who, only a few years before, were regarded as beyond the reach of psychoanalytic treatment.[172]

Among the many psychotherapists inspired by the method and guidance of Sullivan was one who was to leave a legacy of personal influence perhaps as great as his: Frieda Fromm-Reichmann. Her extraordinary success in the treatment of schizophrenia, during the thirties and forties, was a pioneering achievement in more than one sense. Not only did it demonstrate the private world of the "insane" to be accessible to intelligent communication and understanding, but it contributed greatly to the developing reassessment of the therapist's posture in conventional therapy. Her evaluation of that posture, as contrasted to the orthodox view, has been well described by Alexander Gralnick: "At one pole is the reserved and aloof technician; at the other extreme is the sympathetic and active therapist engaged in a life and death struggle to win the patient to his side."[173]

More directly perhaps than any other in her generation, Fromm Reichmann challenged the mystique of transference and countertransference which tended to confine the encounter of therapist and patient to an unreal struggle of shadow-boxing and vicarious role-playing. In the official doctrine the analyst was enjoined to withhold himself as a person, possessing his own values and emotions, entirely apart from the treatment; any departure from this therapeutic neutralism was categorized as "countertransference phenomena"—i.e., as an inappropriate reactivation of his own emotional entanglements in the past.[174] On the other hand *transference* by the patient, in the form of empathic attachments and responses to the analyst, was actively stimulated as an essential step toward solution of his conflict. In time, orthodox Freudians even came to regard the creation of an artificial neurosis, the "transference neurosis," as indispensable to treatment—somewhat on the theory, as Jules Masserman has observed, that the patient must get sicker before he could begin to get better.[175] But this encouragement of transference did not, of course, mean that the patient's gestures toward the doctor were to be granted any face-value meaning or reality content; instead they were ritual re-enactments of infantile dilemmas in which the analyst could figure only as a shadowy surrogate for forgotten objects of the child's love and hate.[176]

There was plainly no room on this darkened stage for a genuine meeting of two real human beings in the present moment. One of the players was encapsulated within his own past (free only to associate), while the other was carefully hidden behind his professional mask (free only to interpret). Any attempt by the analyst to reach out toward the other was forbidden by the rules of the game; any corresponding effort on the part of the patient was disclaimed as illusory. Even the physical setting of the relationship militated against a real confrontation; the analysand lay recumbent on a couch, with the analyst restricted to a chair behind him. With their gaze thus mutually averted, it is not to be wondered that either actor might have difficulty in recognizing the other as a living presence.

Fromm-Reichmann was only one of a host of post-Freudian analysts—among them Schilder, Rado, Alexander, and French, along with Horney and Sullivan[177]—whose invasion of the classical consulting-room resulted in giving it a radically new look. Her contribution differed from the others principally in the degree of its sympathetic warmth and intensity, as well as in the peculiar source of its derivation. As with Sullivan, her therapeutic insights grew out of many years of work with psychotic patients; it is therefore all the more remarkable that they came to be distinguished by deliberate affirmation of such considerations as respect and trust, participation and partnership, the present reality of the patient's problem and his potential capacity to cope with it. In Fromm-Reichmann's system of intensive psychotherapy the analyst ceased to be an aloof authority-figure and became a willing confidante, no longer posing or pretending but committing himself openly and honestly to the mutual quest for enlightenment.[178]

Largely as a consequence of the breakthrough of Sullivan and Fromm-Reichmann in the treatment of schizophrenia, more and more psychotherapists in recent years have found the courage to challenge and transgress long-cherished therapeutic assumptions and conventions. In particular, the strictures surrounding counter-transference—once regarded as an unseemly exhibition of inhibitions—have been not only attacked but virtually turned inside-out as recognition has grown of the positive and constructive value of the *therapist's personality as a therapeutic tool*.[179] Today there is broad agreement that it is undesirable as well as impossible for the doctor to exclude himself from the treatment; that his intervention should in fact be active rather than passive, bold and intimate rather than hesitant and remote—above all, *authentic* rather than devious.[180] In this postmodern view of psychotherapy, the analyst no longer looks down upon the prostrate patient from the heights of Olympus, or looks in curiously upon his conflict from the safe distance of the detached observer. More typically he meets him at close quarters and face to face (the couch is on the way out); he seeks earnestly to share the other's perspective and to join his struggle; while never relinquishing his human feelings and humane connections, he puts aside his textbook preconceptions (as Jung and Rank advised him to do) and willingly suspends his scientific disbelief. In short, as Ludwig Binswanger has expressed it, the psychotherapist "must dare to risk committing his own existence in the struggle for the freedom of his partner's."[181]

To cite Binswanger is to take note of still another stream of contemporary teaching in psychotherapy, emanating from Europe —that which is broadly designated as existentialism. This is not the place for a rounded assessment of either the philosophy or the psychology of this profoundly influential development in man's thought about himself; but its relevance to our immediate interest is direct and unmistakable. Despite the rich diversity of existentialist "schools" and the substantial differences among their creeds, there remains a solid core of agreement about the central message and direction of the movement. It is not too much to say that, in both its philosophical and psychological forms, existentialism provides the most coherent and impressive demonstration available of the humanistic thesis whose evolution we have traced all the way from the physics laboratory to the therapeutic clinic. On one of its sides, to state it anew, that thesis is that man's deepest and most reliable knowledge of the world around him—in particular of himself and others—is attained not through detachment but *att*achment, not by reductive analysis but constructive synthesis, not in a state of estranged aloofness but in something like an act of love. To regard man as an indivisible subject rather than an assembled product—to meet him on his own ground and in his own terms— is, in the existentialist view, not just a moral imperative but a heuristic necessity to be faced by all who seek the truth. There can be no science of man that does not see him whole, as at once organic and humane—both acting and reacting, behaving and intending, determined and determining, being and becoming. In existentialist thought these complementary antinomies of the human condition do not so much conflict as interpenetrate; it becomes more than a play on words to say that freedom is a *necessity* of human existence, and that man is *determined* ("condemned") to be free.[182]

In light of this, it is not surprising that the leading contemporary spokesmen of existentialist philosophy (quite apart from their derivative psychological schools) have been thought to be saying things deeply relevant and fruitful to the concerns of psychoanalysis and psychotherapy. Nor is it much more surprising to find that what they say appears broadly to support and reinforce the findings of the neo-Freudians (as opposed to the pro- or proto-Freudians) and of those psychotherapists who follow in their path. A few examples may be mentioned. The implications for psychotherapy of Martin Buber's dialogical philosophy of "meeting" are considered by his principal American disciple to be parallel in crucial respects to the writings of Erich Fromm and Carl R. Rogers.[183] Another expositor has sought (perhaps more tortuously) to relate Buber to that stream of pragmatic and functional psychology which found its pre-eminent spokesman in G. H. Mead and subsequently shaped the interpersonal perspective of Sullivan and his group.[184]

Again, both Fromm and Sullivan have been favorably compared in their therapeutic aims, by a prominent clinician, to Paul Tillich and his thesis of "healing as participation."[185] Meanwhile the translator and sympathetic interpreter of Sartre, Hazel Barnes, has argued at length that the Sartrean "literature of possibility" is virtually the counterpart in another vocabulary of the analytic psychology of Fromm.[186] And it may be added, finally, that more than one effort has been made to demonstrate affinities linking Adler (with his concepts of holism, social feeling and the life-plan) to the mainstream of existentialism.[187]

None of this activity, to be sure, is very conclusive in itself. The great themes of existentialist thought are sufficiently vast and ambiguous to make plausible, or so it would seem, almost any claim to affiliation. Even Freud has at times been called an "existentialist"; and if that petition is acceptable, surely anything goes.[188] What is of greater weight is the body of writings of the major existentialist philosophers themselves—and, still more in point perhaps, of existentialist psychologists and psychotherapists. Even in this respectable chorus the voices are not in perfect harmony; but on the basis of their testimony it may be possible at least to mark the outer boundaries and keep out the most flagrant trespassers.

The quotation from Binswanger given above (urging that the psychotherapist dare to risk his own existence in encounter with the patient) reveals the influence of Buber, who made use of a similar expression in his preface to a book by the psychoanalyst Hans Trüb, *Heilung aus der Begegnung* ("Healing out of Meeting").[189] Buber's philosophy of dialogue, centering around the I-Thou relation, gives rise directly to a "psychology of meeting" which finds its paradigm in the therapeutic encounter. The significance of the general conception has been ably described by Will Herberg:

> The [term] *I-Thou* points to a relation of person to person, of subject to subject, a relation of reciprocity involving "meeting" or "encounter," while the [term] *I-It* points to a relation of person to thing, of subject to object, involving some form of utilization, domination, or control, even if it is only so-called "objective" knowing. The I-Thou relation, which Buber usually designates as "relation" par excellence, is one in which man can enter only with the whole of his being, as a genuine person.[190]

It follows that the relationship of therapy in its ideal development represents an authentic encounter "on the sharp edge of existence" between two human beings, one seeking and the other helping. This mutual recognition, which is never immediate but only a possibility to be achieved, cuts through the conventional defenses and postures of both partners to permit each to reach out as a person to the other as a person. What is demanded of the doctor in particular is that he "himself step forth out of his protected professional superiority into the elementary situation between one who asks and one who is asked."[191]

In his posthumously published memoirs, Hans Trüb traced

the path of his own discovery and application of Buber's principle of "healing through meeting." In his case the experience involved a painful break with professional associations and doctrinal systems (notably Jung's) which tended to formalize and objectify the clinical relationship and thus to prevent that humane confrontation with the other which, he came to believe, was necessary to evoke the power of healing. Trüb's own existential decision was to risk the loss of professional security, of the comfortable moorings of system and the trappings of respectability, and to venture (in Tillich's phrase) "the courage to be as himself." Of Trüb's life-experiment Buber wrote in his preface:

> This way of frightened pause, of unfrightened deliberation, of personal participation, of rejection of security, of unsparing stepping into relationship, of the bursting of psychologism—this way of vision and of risk is that which Hans Trüb went. Surely other psychotherapists will find the trail that Trüb broke and carry it still further.[192]

Other psychotherapists have indeed found the trail—in ever-increasing numbers. One of the most distinguished among them is Binswanger, leader of the original European school of existential-analysis (*Daseinsanalyse*), whose views of the import of this now-pervasive orientation in Continental psychotherapy are particularly authoritative. The existential viewpoint in psychiatry arose, he writes, "from dissatisfaction with the prevailing efforts to gain a new scientific understanding in psychiatry. . . . Psychology and psychotherapy, as sciences, are admittedly concerned with 'man,' but not at all primarily with mentally *ill* man, but with *man as such*. The new understanding of man, which we owe to Heidegger's analysis of existence, has its basis in the new conception that man is no longer understood in terms of some theory—be it a mechanistic, a biologic or a psychological one. . . ." Whatever may be his formal affiliation or school of thought, the existential analyst "will always stand on the same plane with his patients—the plane of common existence. He will therefore not degrade the patient to an object toward which he is subject, but he will see in him an existential partner."[193]

Another, and wider, application of the I-Thou philosophy to the general field of medicine has been undertaken by a former associate of Buber, the doctor and psychiatrist Viktor von Weizsäcker. In his system of "medical anthropology," all objective diagnoses and specifications are rendered secondary to the task of comprehending the patient as a whole person, a unique and irreducible subject. On the basis of this "inclusive" understanding, Weizsäcker believes, the complementary tasks of objective treatment and technique may proceed without danger of that fragmentation of the person in which his identity is destroyed. "The smooth functioning of the objective practitioner lasts just as long as there is a self-understood relation between doctor and patient, unnoticed because unthreatened. But if the *de facto* assent to this relation falls away, then the objectivity is doubtful and no longer of use."[194]

This complementarity of medical perspectives recalls the observation of Michael Polanyi (noted in Chapter V) that even in organic biology no amount of objective specification or dissection can give the scientist an awareness that the object on the table is, say, a *frog*. Unless that elementary knowledge of the whole organism is given beforehand, the dimension of understanding which it conveys is lacking. With respect to the lesser amphibians and their uses, that information may perhaps be deemed important only to the creature on the table. But in the human use of human beings, to which medical science is thought to contribute, it would seem to be pertinent, if not essential, to proper treatment.

Apart from its incorporation into existentialist schools of psychoanalysis and psychiatry, Buber's immensely fertile concept of I-Thou "meeting" finds parallels and reverberations in various other formulations of existentialist philosophy—notably those of Paul Tillich and Gabriel Marcel.[195] For Marcel, who came independently to the formula of I-and-Thou, the sense of genuine encounter is conveyed by the term "intersubjectivity"—implying an intimate communication on the order of communion.

My experience is in a real communication with other experiences. I cannot be cut off from the one without being cut off from the other. . . . The fact is that we can understand ourselves by starting from the other, or from others, and only by starting from them; . . . it is only in this perspective that a legitimate love of self can be conceived.[196]

This insight, closely similar to Fromm's concept of productive love and self-realization, implies a reciprocity of knowing in which what "*I* am" as well as what "*Thou* art" is made known only through the mutual experience of what "*we* are."[197] Each communicant recognizes the other in himself (to recall Rank's phrase), and recognizes himself in the other.

In Tillich's "therapeutic theology," as it has been called, this general appreciation of the enlightening role of engagement or meeting is applied directly to psychotherapy—regarded as the "community of healing." In common with other existentialists, Tillich believes that the personal troubles represented by neurosis stem fundamentally from failures in relationship with others, resulting in self-alienation from any genuine contact with the world. The central therapeutic problem thus becomes one of "acceptance" —more precisely, of successive stages of acceptance. "In the communion of healing, for example the psychoanalytical situation, the patient participates in the healing power of the helper by whom he is accepted although he feels unacceptable."[198] The patient's acceptance of the helper, supported by the helper's own acceptance of him, leads him gradually to "accept" himself—and thereby his being-in-the-world with others.

In this therapeutic transaction, it should be noted, there are no silent partners; the therapist is no longer a "mute catalyzer" or blank screen but a participant with the whole of his being. He participates, moreover, not only for the purpose of helping but, more basically, for that of knowing or understanding. "You must participate in a self," according to Tillich, "in order to know what it *is*. By participation you change it."[199] The inference is that the kind of knowledge essential to psychology and psychotherapy is to be gained not by detached observation but by *participant-observation* (in Sullivan's phrase). It may be possible, through detachment, to gain knowledge which is "useful"; but only through participation is it possible to gain the knowledge that is *helpful*.

In its broadest meaning this is the counsel, not of subjectivity (and surely not of objectivity), but of *intersubjectivity* as the road to understanding the behavior and experience of the self and of the other. "Existentialism, in short," as Rollo May has put it, "is the endeavor to understand man by cutting below the cleavage between subject and object which has bedevilled Western thought and science since shortly after the Renaissance."[200] It does so not by denying the claims of objectivity in its properly delimited sphere, nor by celebrating the virtues of subjectivity in its monadic purity, but rather by directing commerce to the bridge which stands between. And this "between," if it is not the ground of being, is the ground of *meeting*. "On the far side of the subjective, on this side of the objective," Buber points out, "on the narrow ridge, where *I* and *Thou* meet, there is the realm of 'between.' "[201]

Man's freedom of choice—the basis of his existential indeterminacy and personal responsibility—is seen to be confirmed by the several convergent traditions in postmodern psychology which the present chapter has surveyed. The cumulative evidence from psychobiology, perception, psychoanalysis, psychotherapy and existential psychology therefore gives us the groundwork for a new and reconstructive science of human behavior—an alternative to "behavioral science." The main outlines of that science are now identifiable.

It will be a science whose guiding purpose is not the measurement of organic mechanisms or the manipulation of conditioned responses but the understanding of personal experience in its complementary wholeness: a science which, in Riezler's words, begins with "respect for the subject-matter"—and ends in vindication of that respect.

It will be a science activated, not by a rage for order, but by a passion for freedom.

It will be a science which regards men as actors as well as spectators, and accordingly perceives its own task as one primarily of participation (intersubjectivity) and only secondarily of observation (objectivity). In short, it will recognize with Tillich that "detachment is only one element within the embracing act of cognitive participation."[202]

It will be a science which, in seeking to comprehend human nature and conduct, takes men's reasons and reasoning into account as seriously as it does nature's causes; one that makes of its inquiry into truth and understanding a reasonable dialogue, in which the other partner (the observed) has an equal right to be heard and even to be trusted—not because this is a generous thing to do, but because it is indispensable to the inquiry. For the inquiry of this humane science will be predicated on the outrageous hypothesis, as proposed by the author of *The Organization Man,* "that people often do what they do for the reasons they think they do."

The constructive science of behavior will dare to look upon all men as moral agents, and upon their behavior as the expression of a choice—in agreement with Sartre that "this decision is human, and I shall carry the entire responsibility for it."[203]

And it will agree with Socrates, in his final words to his scientific friends, that even in the last extremity the mind of man has reasons that his organic machinery knows not of:

> For, by the Dog! these bones and sinews, I think, would have been somewhere near Megara and Boeotia long ago, carried there by an opinion of what is best, if I had not believed it better and more just to submit to any sentence which my city gives than to take to my heels and run. But to call such things causes is strange indeed. If one should say that unless I had such things, bones and sinews and all the rest I have, I should not have been able to do what I thought best, that would be true; but to say that these, and not my choice of the best, are the causes of my doing what I do . . . would be a very far-fetched and slovenly way of speaking.[204]

CHAPTER VII

The Human Image:
Science and the Understanding of Man

ONE OF THE MOST INSTRUCTIVE ANALOGIES in the history of ideas is that which the late R. G. Collingwood found to lie at the heart of the Idea of Nature—as reflected, more particularly, in the strange career of the concept of "cause."[1] The notion of causality in nature, as he reminded us, arose originally out of direct analogy with the *self*: that is, with the inner resources of will and purpose guiding the conduct of a free and responsible human being. The early Greeks, taking note of the operations of their own minds, simply projected these outward upon the world—thereby making of nature an "intelligent organism." In short, the object was regarded as a subject; the natural world was anthropomorphized.[2]

Later on, as we know, with the advent of modern science, nature came to be shorn of its intelligence and organicism in favor of a more congenial analogy: that of the machine. For a time, during the reign of Cartesianism, an uneasy balance of power prevailed between the dual realms of *res cogitans* and *res extensa*. But the momentum of the belief in matter, once set in motion, was not so easily confined; it soon overflowed the boundaries, and in the process the anthropomorphic analogy was exactly reversed. The human self, originally the source of explanation for the workings of inanimate nature, came in the end to be explained in the mechanical terms of natural causation. The subject was now regarded as an object; the human world was mechanomorphized.

It has been a thesis of this book (specifically of Part One) that that is still predominantly the state of affairs in contemporary institutes of behavioral science, whose teams of technicians are conscientiously at work perfecting the naturalization of the human subject. The fundamental premise on which this program rests may be briefly rehearsed by means of a representative quotation, from a work entitled *The Scientific Study of Social Behaviour*. In the process of demolishing an unacceptable theory of psychology (the phenomenological approach of Snygg and Combs) which seeks to take account of inner experience as well as of overt behavior, the author observes:

> The first general comment on this approach concerns the philosophical position which it assumes. It is supposed that conscious events cause the behaviour which they accompany. This is no place for metaphysical arguments, but it may be said that many philosophers would be in disagreement with this position, on the grounds that, whatever conscious events are, they are not the sort of things which can be regarded as causing behaviour —they fall into the wrong logical category.[3]

What is noteworthy about this passage is its confident acceptance of a doctrine of causality which, with respect to human conduct, requires at the outset the *categorical* exclusion of conscious experience. "Conscious events" are no longer to be regarded even as relevant variables deserving of investigation; they simply "fall into the wrong logical category," and that is that. On this view it would not seem too much to say (paraphrasing slightly a remark of Skinner) that the hypothesis that man is *not conscious* is essential to the scientific study of behavior.

Let us see where this line of explanation leads. A contemporary philosopher of politics, Joseph Tussman, has taken note of the fact that our theories of political behavior, with their focus upon objective description and prediction, are invariably written from the standpoint of an outside observer. Because of this, he points out, they are helpless to assist the *behaver* in deciding *how he should act*. But that is, of course, just his problem. Once we recognize the object of behavioral theory to be himself an active agent, a decision-maker, we acknowledge a perspective for which our predictive theory is irrelevant. For instance: "The judge, manning the judicial tribunal, is confronted with a problem. He is not trying to predict his own behavior; he is concerned with behaving properly, with determining what he 'should' do. And that is quite another matter."[4] The judge's task or responsibility, according to Tussman, can be accommodated only by a normative Theory of Political Obligation—which exists in another realm of discourse than that of a descriptive Theory of Political Behavior. If we are to help the judge make a proper choice, in other words, we need a literature of social science appropriate to decision-making.

Unfortunately, Tussman does not go on to draw the conclusion that an accurate *prediction* of the judge's behavior (i.e., of his decision) must therefore take into account the Theory of Obligation. It would seem that, if the judge does in fact decide on these grounds, they become the grounds for explanation and anticipation of his conduct. Tussman is satisfied to distinguish between what he regards as the separate concerns of "prediction and performance"—of objective analysis and subjective choice— and to grant them each a sphere of validity. That is itself, no doubt, a judicious decision. But if we accept the validity of the realm of choice at all, how can we deny its relevance to a predictive theory which seeks to discover what the judge will do?

We might make another, perhaps fanciful, supposition about the judge. Let us imagine him to have been trained in the orthodox School of Political Behavior, prior to his appointment to the bench, and moreover to be thoroughly disenchanted with all metaphysical notions of political obligation. How then will he go about making his decision? Presumably he will try to *forecast* it (like Caesar in a famous moment of paralysis of the will) by consulting the resident augurers and soothsayers. These specialists in behavioral prediction will not be able to offer him a perfectly sure thing; but they will skillfully estimate the probabilities and hand down the odds. The judge will take account of statistical trends in judicial behavior, as reported in the *American Political Science Review*,[5] as well as of the role-expectations that significant interest-groups have of him. Having calculated all the variables and opportunities, he will then act "rationally"—which is to say, *opportunistically*.

How could he do otherwise? He has not been encouraged to place any credence in his conscious reason or capacity for independent judgment. He knows that the still small voice which murmurs to him of responsibility and choice is only a disease of the inner ear—and that, "whatever conscious events are, they are not the sort of things which can be regarded as causing [his] behavior." If he is not to be laughed out of court, the judge must look upon himself and his role objectively, from the outside; and in so doing he becomes, in the fullest sense of the word, irresponsible. He has refused to judge.

This brings us to another irony in the history of thought, not unrelated to the idea of causality. One of the most elusive, and delusive, terms in the entire vocabulary of social science is that of "rationality." In its classical definition, rationality was the distinguishing characteristic of human beings; man was the rational or thinking animal, *Homo sapiens*. But the triumph of modern science carried with it a purification of the idea of reason, first through the doctrine of rationalism and later that of empiricism. "Reason" for the Enlightenment was not merely the antonym of "faith"; it was equally opposed to all feeling and sentiment, sympathy and intuition. (If it had been otherwise, there would have been no grounds for the Romanticist revolt.) But these affective sensibilities are obvious ingredients of the human condition; man in his ordinary life and relations, his social institutions and common language, is conspicuously entangled in affection and bias. It was therefore recognized by the more consistent rationalists that very little "rationality" was to be found in the run of humanity; on the other hand it was evident how much of it there was to be found in the logical operations and symmetrical structures of science. The concept of rationality, thus redefined on the model of science and technology, receded ever farther from its original human reference point—until at last it took up residence in the machine.[6]

This course of events is graphically mirrored in the familiar technological concept of "rationalization." The typical processes of industry and commerce, of large-scale organization and its formal administration, are regarded as *rationalized* to the extent that they have been freed from the frictions of disturbance and distraction—the extent, that is, to which they approximate the norm of mechanical efficiency. It follows (and indeed is often admitted) that, to the degree to which they "act like human beings," these processes are *irrational*.[7]

But this historic displacement of rationality from man to the machine is not yet the end of the story. The final step was taken with the attempt to import rationality back into the deflated image of man by defining his behavior strictly in terms of the reigning mechanical model. We have seen in Chapter 1 how this effort inspired the construction of such conceits as that of Economic Man (alias "Rational Man")—the organic cash-register or pleasure-machine, endlessly totting up the balance of profit and loss, of pleasures and pains, by means of the felicific calculus of rational self-interest.

The obituary of the classical model—the end of Economic Man—has been fully reported by Peter F. Drucker and others.[8] It is enough to recollect here that the theoretical model broke down completely when confronted with the elementary facts of human conduct, whether in the marketplace or in the forum. But what is more remarkable is that the failure of the mechanical-rational model to account for the facts of behavior has commonly been taken to mean, not that the *model is unreasonable*, but that *human behavior is irrational*. In effect it was not the behavioral theory that broke down, but man himself. When he was seen to fall short of the lofty expectations set for him—when he failed to perform consistently at the level of an omniscient and omnicompetent robot—the conclusion had to be that man's conduct was irrational, irresponsible, and capricious.

From the standpoint of the scientific observer, with his

rarefied doctrine of perfect rationality, this pessimistic conclusion was doubtless unavoidable. But (as we have observed at some length) that standpoint, although it dominates the behavioral sciences, no longer holds the field alone. It is directly challenged by a contrasting opinion which asserts that men on occasion do act from choice as well as from compulsion; that their decisions may in fact be guided by conscious purposes as well as driven by involuntary causes. What is more to the point, this competing doctrine holds that the *meaning* of any human action—including whether it may be judged as reasoned or reasonable in the circumstances—cannot be known otherwise than by awareness of the peculiar constellation of felt needs and interests, the world of personal experience, within which the action has its source and relevance.

A cogent illustration of this perspective at work is to be found in a series of studies conducted in the 1950's by the Yale Labor-Management Center, dealing with such questions as job satisfaction and mobility among industrial workers. In one monograph of the series, the authors (Lloyd G. Reynolds and Joseph Schister) point out that the thinking of management has been dominated by the conventional theory of job choices which emphasizes "rationality and economic motivation." According to this conception "the worker is regarded as behaving like a scientist, carefully gathering all of the relevant facts, and then choosing the job which promises the greatest net advantage." The authors' own study indeed confirms the hypothesis that a wide gap exists between the theory and the actual facts of behavior; but their inference from that finding is not the usual one. "This does not entitle us, however, to say that the observed behavior is irrational. We should ask ourselves whether it may not be economic theory which is unreasonable." And the thesis which they proceed to develop and document is that "worker behavior is in general a rational adaptation to the circumstances *as the worker sees them.*"[9]

What this theory argues is that we cannot understand the behavior of a human being in strictly "behavioral" terms. If we wish to know the meaning of behavior, we must know the *mean-*

ings of the behaver; to remain outside his frame of experience is simply to remain in the dark. (It may be recalled that this was very much the discovery which overtook the perceptual psychologists Solley and Murphy, at the conclusion of experiments in which they had remained studiously at "the simple observational level.")[10] For despite the behaviorists of psychology, as Dewey long ago insisted, behavior is not antecedent to experience but consummatory; it is not the cause but the expression of purpose and value. And despite the behavioralists of social science, behavior is not a discrete event or independent variable to be caught and studied while all else is held constant; like the electron of de Broglie, it cannot be forcibly separated from the self-system of which it is a part without doing violence both to the system and to the part.

That old urge in all of us (which Sherrington remarked upon) to get away from the bogey of "subjective anthropomorphism" acquired perhaps its greatest reinforcement of recent decades through the method of "operational analysis" introduced by the mathematical physicist Percy W. Bridgman.[11] In the perspective of operationalism, the meaning of a proposition is simply the particular set of "operations" required to implement or verify it. Bridgman's concept had the virtue of seeming to fulfill at a stroke the positivist yearning for absolute objectification—to get rid once and for all of that brooding omnipresence, the human observer, with his trembling fingers and contaminating biases. In the years following publication of *The Logic of Modern Physics* in 1928, numbers of scientific philosophers and scientific psychologists embraced Bridgman's method and proceeded to accord it the full degree of reverence to which Rollo May has given the name of methodolatry.

But meanwhile a curious thing was happening. The author of operationalism had himself begun to reappraise and revise his concept in the light of momentous developments in his own fields of physics and mathematics—mainly those associated with Heisenberg's quantum principle and Gödel's logical theorem.[12] And the conclusion to which Bridgman found himself more and more drawn was "the insight that we never get away from ourselves"—

that the operations involved in any scientific performance, or for that matter any human act, are ultimately and irreducibly *individual*.[13] There was, he concluded, simply no escaping the personal reference point, the stubborn particularity of the solitary knower doing his job. In time this heretical line of thought led Bridgman to an appreciation of "subjective anthropomorphism" so profound and unequivocal that he was compelled repeatedly to defend himself against the dread charge of solipsism—leveled at him in particular by his own erstwhile disciples, the operationalists.[14] It might be said that Bridgman, having extracted meaning and relevance from the human subject through an ingenious feat of surgery, spent the balance of his career trying to put them back. In the process he became an outspoken advocate of the thesis of personal involvement—of the reality of participation and the importance of concern—in the behavior both of men in the street and of scientists in the tower.[15]

This recognition that, in Polanyi's phrase, "knowledge is personal"—and most profoundly when it is knowledge of other persons—has been widely resisted on the assumption that it constitutes a confession of human frailty and therefore (as Hull considered it) a counsel of despair. And no doubt to those who are convinced that the stockpile of human knowledge increases in direct ratio to the distance between subject and object, the very notion of "personal knowledge" can have the appearance only of a blooming, buzzing confusion of logic. But it is instructive to be reminded, once more, of the singular fact that throughout three centuries of scientific ascendancy the lingering suspicion that we cannot get away from ourselves—and even, perhaps, that we should not—has never been altogether silenced or suppressed. Phrased another way, this is the view which holds that to acknowledge the the inner personal dimension in behavior, as well as in the *observation* of behavior, is not to confess a fatal defect but to identify a source of strength—a unique prerogative of the human condition which provides the sole basis for genuine communication between man and man, and hence for the understanding of other voices, other minds and other behaviors.[16]

The method of "understanding" (*Verstehen*) which this viewpoint makes central to the study of mankind is, of course, far from novel. It is at least as old as the philosophical dialogue of Socrates—and as current as the dialogical philosophy of Buber.[17] In the last century it found its most effective formulation in the writings of Wilhelm Dilthey, which exerted a pervasive influence upon the development not only of existentialism and phenomenology but of the mainstream of European social science.[18] For Dilthey the essential, and unbridgeable, distinction between the natural sciences (*Naturwissenschaften*) and the cultural or human sciences (*Geisteswissenschaften*) was to be seen in the difference between *causal processes*, objectively perceived, and *meaningful relations* disclosed to us through our innate capacity for understanding. Life calls out to life, he insisted: we can make sense of another's actions because we ourselves are actors; we can comprehend his reasons only because we know what it is like to reason.[19] To the extent that we "know ourselves," through introspection, we are equipped to know another—through what Cooley termed "sympathetic introspection." And to the extent that we discern the meaning of his lived experience we come to apprehend the meaning of our own. "Understanding," said Dilthey, "is the rediscovery of the I in the Thou."[20]

Dilthey's hermeneutic science of experience and understanding was a seminal force at the turn of the century in the remarkable (if incomplete) emancipation of European social thought from the bonds of that scientific world view whose triumphal career we have traced in the opening chapter of this book. Most directly of all, as a consequence of his own deep concern with the problem of the past, Dilthey's vision gave impetus to the new school of "historical understanding" which—at the hands most notably of Croce, Trevelyan and Collingwood—set itself firmly against the "scientificizing" tendency threatening to preempt the field of historiography.[21] In their forceful and cumulative insistence that Clio is not a mechanism but a muse; that (in Mommsen's phrase) great history is neither written nor made except through passionate participation; that such things as imagination, viewpoint, sympathy and commitment are indispensable attributes of the historian's craft—in all of this the anti-scientific school strove mightily to redress the balance of scholarly trade with the past. Largely as a result of their labors the discipline of history has, to a degree unique among the social studies, succeeded in remaining recalcitrant (although scarcely immune) to all attempts to make it over into a positively behavioral science.[22]

At the same time, Dilthey's influence and example were apparent in the significant departures taken by European psychology and sociology in the early years of the century. A broad movement was then under way to construct a "philosophical anthropology," a new image of man, which would accord with discoveries in the life sciences as well as with the spreading conviction that conventional scientific analysis had reached the point of diminishing returns.[23] The shifting groundwork of social inquiry and self-inquiry is indicated by the observation of Max Scheler that man, for the first time, had become problematic to himself. As much as anyone else of his generation it was the redoubtable Scheler—with Husserl a pioneer of phenomenology, with Bergson a champion of intuition and creative evolutionism, with Mannheim a founder of the sociology of knowledge—who fruitfully explored and traversed the new pathways opened up by the principle of *Verstehen*.[24] In *The Nature of Sympathy*, most explicitly, he argued for the return to respectability of that natural resource of humane communication which centuries of scientific infatuation had stripped of all dignity and standing.[25] For Scheler the quality of sympathy—and still more its apotheosis of love—was no mystical or romantic effluvium of interest only to poets: it was rather a reliable instrument of knowledge, a means to rational understanding of the human condition in all of the dimensions of direct concern to social science.

No contemporary social scientist who has not completely turned his back upon what C. Wright Mills termed the "classic tradition" in social thought can afford to dismiss or disregard the central place occupied in that tradition—not only in Europe but in America as well—by the method known as *Verstehen*.[26] Indeed it would require a separate, and sizeable, volume to do descriptive justice to the main currents and minor tributaries of this conceptual approach. The greatest of its exponents on the continent was, of course, Max Weber; its greatest advocate in America was William James.[27] Although the towering stature of these two intellectual giants has not since been equalled, their insights and achievements have inspired entire schools of social science and philosophy—not a few of which are still vigorously alive in our own day.

Among them is the school of thought associated principally with the names of Scheler and Mannheim: that which has become known as the sociology of knowledge. Because the impact of this provocative movement is still being acutely felt and argued, it is in point to take note briefly of its basic methodological premise. As forthrightly stated by Louis Wirth in his introduction to Mannheim's *Ideology and Utopia*, that premise is grounded in the familiar distinction between the spheres of physical science and social science, and in "the corresponding differentiation between the modes of knowing these two kinds of phenomena." On the one hand physical objects "can be known purely from the outside, while mental and social processes can be known only from the inside"—as well as, on occasion, from the meanings which we read into their physical indexes. *"Hence insight may be regarded as the core of social knowledge. It is arrived at by being on the inside of the phenomenon to be observed. . . . It is the participation in an activity that generates interest, purpose, point of view, value, meaning, and intelligibility, as well as bias."*[28]

Mannheim's own deepest animus, on the methodological front, was directed against the incursions of Watsonian behaviorism in the social sciences. With its positivist urge to reduce everything to "a measurable or inventory-like describability," this tendency was seen to threaten a "mechanistic dehumanization" of the whole range of social experience. Mannheim's most telling point against the behaviorist approach was that "no real penetration into social reality" was possible on its terms—precisely because it refused to penetrate beneath the surface of external behavior and systematically excluded "every significant formulation of a problem." No human situation or social event, he argued, could be remotely comprehended through the superficial cues and appearances which the behaviorist fastened upon. On the contrary, "a human situation is characterizable only when one has also taken into account those conceptions which the participants have of it, how they experience their tensions in this situation and how they react to the tensions so conceived."[29]

Not only were these subjective "definitions of the situation" (to use the phrase of W. I. Thomas) indispensable to an adequate assessment of its meaning; they could be grasped by the mind with a depth and clarity far surpassing the degree of knowledge derivable from "strictly external formalized elements." Following Dilthey (and anticipating the later existentialists), Mannheim appealed to social scientists in effect to "risk themselves" —that is, to risk the possibility of bias and dogmatism in their science—by an act of open commitment, of genuine participation in the stream of human activity they were concerned to explain. The depths of that stream were not to be plumbed by judicious soundings taken from a safe shoreline; they could be fully known only through the baptism of immersion. For Mannheim no less than for Croce, the understanding of human events involved an "imaginative reconstruction" of the situation as experienced—and this in turn depended upon the extent of the scholar's own engagement and concern with the humanity around him.[30] "Indeed, . . . participation in the living context of social life is a presupposition of the understanding of the inner nature of this living context."[31]

It is noteworthy that, in his methodological arguments, Mannheim was far from seeking to disparage "objectivity" in the name of a radical or ineffable "subjectivity," such as Romanticism had undertaken. What he sought was rather to redefine for social science the fundamental relationship between subject and object which the standard canons of the field had ordained as one of absolute detachment and disinterest. Where for the orthodox behaviorist the distance between observer and observed was a vast Newtonian void filled with the ether of indifference, the sociologist of knowledge proceeded on the basis of an alternative theory of "relativity," or relationship, which undertook to bridge the gap (and clear the air) by abolishing the ether. The main point was that, with respect to things human, it is not disinterest that makes knowledge possible but its opposite; without the factor of *interest*, in the primary sense of concern or care, there can be no recognition of the subject matter in its distinctive human character—and hence no real awareness of its situation and no understanding of its behavior.[32]

In the development of his sociology of knowledge, Mannheim was very much aware of the quality of *complementarity*— the alternation of objective and humane perspectives—in the study of human affairs. In confronting the question, "Is a science of politics possible?", he followed his Austrian colleague Albert Schäffle in distinguishing two separate aspects or dimensions of political life which answer to two different scientific approaches. The first is the realm of recurrent and determinate events, those that follow conventional paths and prescribed patterns—which Schäffle had designated as "the routine affairs of state." The second (for which he reserved the title of "politics") consists of the ongoing process of unique and indeterminate occurrences—"the state and society in so far as they are still in the process of becoming."[33] For Mannheim it was the latter realm with which a science of politics was properly concerned, and so he rephrased the question to read: "Is there a science of this becoming, a science of creative activity?" Such a science would accept the ambiguity of a subject matter pervaded by uncertainty and alive with spontaneity; abandoning the effort to corral its free spirit within the iron cage of mechanistic analysis, the new science of politics would affirm its own creative task (its "purposefully oriented will") by affirming the freedom and creativity of its subject matter.

In a striking passage toward the end of his book, Mannheim declared his own position with respect to the "rationalization" of society and the reduction of the human enterprise to the predictive rounds of conditioned behavior. "It is possible," he wrote,

> that in the future, in a world in which there is never anything new, in which all is finished and each moment is a repetition of the past, there can exist a condition in which thought will be utterly devoid of all ideological and utopian elements. . . . The

disappearance of utopia brings about a static state of affairs in which man himself becomes no more than a thing. We would be faced then with the greatest paradox imaginable, namely, that man, who has achieved the highest degree of rational mastery of existence, left without any ideals, becomes a mere creature of impulses. Thus, after a long, tortuous but heroic development, just at the highest stage of awareness, when history is ceasing to be blind fate, and is becoming more and more man's own creation, with the relinquishment of utopias, man would lose his will to shape history and therewith his ability to understand it.[34]

In introducing Mannheim's thought to the English-speaking world, Louis Wirth pointed to its close connections with the pragmatic school in American sociology and social psychology—in particular the work of James, Mead, Dewey, Cooley and Thomas.[35] While so broad a canvass is undoubtedly precarious at some points—Dewey, for example, rather like Weber, was also drawn on one of his sides to a position adjacent to positivism—the generalization serves to underscore the definite and sustained presence of the method of "understanding" in the American wing of the classic tradition. It was Cooley who gave currency to the precept of "sympathetic introspection"; and it was James who emphasized the qualitative distinction between objective "knowledge about" and intersubjective "acquaintance with."[36] We have already seen elsewhere that at least one student has found it possible to interpret Mead's symbolic interactionism as a kind of pragmatic equivalent of the dialogical *Verstehen* of Martin Buber.[37] (And it may be noted again that Mead's psychoanalytic disciple, Sullivan, gave renewed vitality and application to the same principle through his therapeutic posture of participant-observation.)

The recognition of this decidedly "anthropomorphic" perspective in the study of human nature and conduct has not, of course, been the exclusive prerogative of the pragmatists. In America as in Europe, the disciples of idealism (along with intuitionists and Romanticists) have all along been making their own distinctive uses of it. Thus, for a notable example, in Hocking's tolerant and sophisticated philosophy of "objective idealism," the two dimensions of human existence (and the two ways of encountering it) which we have called the *organic* and the *humane* are reconciled within a conception of the self as "a union of opposites"—at once a thing of nature and more than a thing of nature. Man as organism is determinate and specifiable, an object (it) for science; but man as man is *free*, a subject (I) to himself and a subject (Thou) for others.[38] The means of knowing, in force between man and man—and indeed between man and the universe—are as inclusive as the range of human communication. "And at this point," writes Hocking, "our experiences of love and beauty have a decisive word to say. We speak of them as 'feelings': what if they are also *knowings?* I suggest that they are such: that they afford an initiation into the nature of objective reality—in brief that they are, of themselves, not emotions only but moments of metaphysical insight."[39]

It will be apparent how close this is to the great philosopher-scientist of the modern world, Alfred North Whitehead, who progressed over a long career from the rational abstractions of mathematical physics to the construction of a world view in which *feeling*, rescued from the underworld, was elevated into the position of first principle. It is well-known that Whitehead was influenced in his ultimate philosophy by the theory of quantum physics; perhaps it was not entirely coincidence that he was brought to an intuition very like Bohr's of the coexistence of two complementary "modes of understanding"—whether that understanding be of nature, ourselves, or the universe. "The first mode may be called the internal understanding, and the second mode is the external understanding. . . . The two modes are reciprocal: either presupposes the other. The first mode conceives the thing as an outcome, the second mode conceives it as a causal factor."[40] The causal mode of external understanding, Whitehead maintained, had thoroughly dominated the stage of modern thought and swept aside as unintelligible every effort to modify or correct it. But in fact the scientific ground was largely gone from under its pretensions; it had never truly sufficed for more than partial and pragmatic understandings, and even those were now diminished by the progress of the new physics. Nor was this limitation of vision a merely academic issue. "If civilization is to survive, the expansion of understanding is a prime necessity."[41] And the direction which that expanded understanding must take was made clear by Whitehead in a paragraph of his *Modes of Thought* which is worth quoting in full.

> My aim in these lectures is briefly to point out how both Newton's contribution and Hume's contribution are, each in their way, gravely defective. They are right as far as they go. But they omit those aspects of the Universe as experienced, and of our modes of experiencing, which jointly lead to the more penetrating ways of understanding. In the recent situations at Washington, D.C. [1933], the Hume-Newton modes of thought can only discern a complex transition of sensa, and an entangled locomotion of molecules, while the deepest intuition of the whole world discerns the President of the United States inaugurating a new chapter in the history of mankind. In such ways the Hume-Newton interpretation omits our intuitive modes of understanding.[42]

On this view, then, it is the intuitive modes of thought and observation—those which bring into focus the world as experienced—which "lead to the more penetrating ways of understanding." By their aid we are enabled to discern the human purpose and pattern in what otherwise must remain "an entangled locomotion of molecules." But it is exactly these modes of internal understanding, these doors to the Within of things, which have been systematically sealed off by students of behavior dedicated to the "Hume-Newton" model. Their heuristic model does, to be sure, also furnish a framework of explanation for the molecular locomotion in Washington; it is to be studied there as it is studied in the laboratory, by minutely tracing the involuntary responses of statesmen to their objective causal stimuli. For that is all that this "behavior," taken by itself, can ever exhibit to the gaze of the outside observer—and, in any event, that is all he is determined to see.[43]

We have observed at length, in the first half of the present book, the diminishing returns of this investment in behaviorism: the shriveling consequences for the self-image of man imposed by its one-eyed perspective upon his nature and conduct. But we have also discovered, in Part Two, an alternative and compensating mode of perception which proposes to correct the bias of "mecha-nomorphism" and restore normal vision to the social scientist. In brief, the outlook of the observer requires to be balanced by the insight of the agent.[44] When that is done, more is changed than the angle of vision: the object disappears and is replaced by a subject. What then emerges into view is none other than the forgotten man of contemporary behavioral science: the missing human person. We may now appreciate the full significance of that "immense evocative analogy" inspired by the principle of complementarity. When man is the subject, the proper understanding of science leads unmistakably to the science of understanding.

Notes and References

Part One

Chapter I *The Mechanization of Man*

1. J. Robert Oppenheimer, *Science and the Common Understanding* (New York: Simon and Schuster, 1954), pp. 13–14.

2. "In this so-called classical mechanics all reference to purpose is eliminated, since the course of events is described as automatic consequences of given initial conditions." Niels Bohr, *Atomic Physics and Human Knowledge* (New York: Wiley, 1958), p. 95. Cf. E. A. Burtt, *The Metaphysical Foundations of Modern Science* (Garden City, N. Y.: Doubleday Anchor Books, 1955), p. 104; John Herman Randall, Jr., *The Making of the Modern Mind* (New York: Houghton Mifflin, rev. ed., 1940), p. 259.

3. Quoted in Alexander Koyré, *From the Closed World to the Infinite Universe* (New York: Harper Torchbooks, 1958), p. 29.

4. Someone must surely have remarked that it was another Galilean, sixteen centuries before, who had first placed man at the center of the universe.

5. *Il Saggiatore* (Bologna, 1655), pp. 150–163; quoted in J. W. N. Sullivan, *The Limitations of Science* (New York: Viking Press, 1933), p. 130. The idea of "universal mechanism," as Crombie points out, had begun to take hold well in advance of Galileo. Bacon, most notably, "was one of the earliest modern writers to propose the complete reduction of all events to matter and motion." A. C. Crombie, *Medieval and Early Modern Science* (Garden City, N. Y.: Doubleday Anchor Books, 1959), Vol. II, p. 290. On Galileo's method, see A. R. Hall, *The Scientific Revolution, 1500–1800* (Boston: Beacon Press, 1956), pp. 168–178.

6. "After the close of the seventeenth century," according to Whitehead, "science took charge of the materialistic nature, and philosophy took charge of the cogitating minds." Philosophy ever after "has retreated into the subjectivist sphere of mind, by reason of its expulsion from the objectivist sphere of matter." *Science and the Modern World* (New York: Mentor Books, 1948 [original ed. 1925]), pp. 145, 142. But this overlooks the fact that philosophy has often refused even to defend the final bastion of the "subjectivist" sphere, and, in its various materialist and positivist forms, has appeared to outdo science itself in rejection of the cogitating mind.

263

7. *The Metaphysical Foundations of Modern Science*, pp. 238–239.

8. *From the Closed World to the Infinite Universe*, p. 4.

9. Crombie, *Medieval and Early Modern Science*, p. 294. Pope's famous couplet is entitled "Epitaph, intended for Newton's tomb in Westminster Abbey."

10. E. N. da C. Andrade, *An Approach to Modern Physics* (Garden City, N. Y.: Doubleday Anchor Books, 1957), p. 259. Compare the shrewd description of primitive magic by Sir James Frazer: "Thus its fundamental conception is identical with that of modern science; underlying the whole system is a faith, implicit but real and firm, in the order and uniformity of nature. The magician does not doubt that the same causes will always produce the same effects, . . . Thus the analogy between the magical and scientific conceptions of the world is close. In both of them the succession of events is assumed to be perfectly regular and certain, being determined by immutable laws, the operation of which can be foreseen and calculated precisely; the elements of caprice, of chance, and of accident are banished from the course of nature. Both of them open up a seemingly boundless vista of possibilities to him who knows the causes of things and can touch the secret springs that set in motion the vast and intricate mechanism of the world." *The Golden Bough: A Study in Magic and Religion* (New York: Macmillan, 1951, one-vol. ed.), p. 56.

11. Jacques Maritain, *The Dream of Descartes* (New York: Philosophical Library, 1944), p. 21. Cf. S. F. Mason, *Main Currents of Scientific Thought* (New York: Abelard-Schuman, 1956), p. 135.

12. *The Making of the Modern Mind*, pp. 241, 242. Maritain further interprets the philosopher's dream as follows: "It is in the unity of the *admirable science*, which will imply one single light and single mode of certitude . . . [for which] he must put aside the traditional idea of the specific diversity of the sciences, a singularly grave decision whose effects are not yet fully realized. What Descartes is entrusted to give to men is a 'universal science' embracing all things knowable in his perfect specific unity—THE SCIENCE." *The Dream of Descartes*, p. 25.

13. Burtt, *The Metaphysical Foundations of Modern Science*, p. 126.

14. *Ibid*. Cf. Hobbes's introduction to the *Leviathan*: "For seeing life is but a motion of limbs, the beginning whereof is in some principal part within; why may we not say, that all *automata* (engines that move themselves by springs and wheels as doth a watch) have an artificial life? For what is the *heart*, but a spring; and the *nerves*, but so many strings; and the *joints*, but so many *wheels*, giving motion to the whole body, such as was intended by the artificer?" *Leviathan* (New York: Oxford University Press, 1947), p. 5.

15. Cf. Stuart Hampshire (ed.), *The Age of Reason* (New York: Mentor Books, 1956), p. 36.

16. "Hobbes applied his idea of the basic nature of motion to the world of human conduct. His politics were derived from a psychology that he based in turn on physics. The whole was rigorously developed in what Hobbes hoped, quite wrongly, was a demonstration, *more geometrico*, of politics as a branch of physical science." William Y. Elliott and Neil A. McDonald, *Western Political Heritage* (New York: Prentice-Hall, 1950), p. 431.

17. Charles Frankel, *The Faith of Reason: The Idea of Progress in the French Enlightenment* (New York: King's Crown Press, 1948), p. 14.

18. *The Philosophical Works of Leibnitz*, trans. George Martin Duncan (New Haven: 2nd ed., 1908), p. 269. Cf. Maritain, *The Dream of Descartes*, p. 96.

19. Randall, *The Making of the Modern Mind*, pp. 246–247.

20. On the present-day controversy provoked by the quantum revolution, see below, Chap. IV.

21. *Ethics*, quoted in Randall, *The Making of the Modern Mind*, p. 247.

22. See, for example, Hampshire, *The Age of Reason*, pp. 100–101.

23. Feuer points out that in Spinoza's time "the rise and growing vogue of the mathematical method made the conception of determinism seem axiomatic. Whenever a new 'method' has been proposed in the history of thought, its converts and proponents have seen in it the promise of a new age. . . . Every method has had its methodological madness, and among the young Cartesians, of the left and right wings, in the seventeenth century, the mathematical method augured tremendous conquests." *Spinoza and the Rise of Liberalism*, p. 240.

24. *Ibid.*, pp. 76–77.

25. *Ibid.*, p. 77.

26. *Ibid.*, p. 242.

27. Quoted by Feuer, *ibid.*, p. 243.

28. *The Condition of Man* (New York: Harcourt, Brace, 1944), p. 176.

29. R. H. Tawney, *Religion and the Rise of Capitalism* (New York: Penguin Books, 1947 [original ed. 1926]), p. 17.

30. *Ibid.* Cf. Werner Sombart, *The Quintessence of Capitalism* (New York: Dutton, 1915), chap. VIII, "The Art of Calculation," pp. 125 ff.

31. Laplace, *Traité de Probabilité* (1886); quoted in Michael Polanyi, *Personal Knowledge: Towards a Post-Critical Philosophy* (Chicago: University of Chicago Press, 1958), p. 140. Polanyi, after a mathematical demonstration of "the stark absurdity of this claim," goes on to point the moral in trenchant terms: "Yet the spell of the Laplacean delu-

sion remains unbroken to this day. The ideal of strictly objective knowledge, paradigmatically formulated by Laplace, continues to sustain a universal tendency to enhance the observational accuracy and systematic precision of science, at the expense of its bearing on its subject matter. . . . I mention it here only as an intermediate stage in a wider intellectual disorder: namely the menace to all cultural values, including those of science, by an acceptance of a conception of man derived from a Laplacean ideal of knowledge and by the conduct of human affairs in the light of such a conception." *Ibid.*, p. 141.

32. Thus Cassirer has summarized the viewpoint: "If we remove the mask of words, of arbitrary concepts, of phantastic prejudices from the face of nature, it will reveal itself as it really is, as an organic whole, self-supporting and self-explanatory." Ernst Cassirer, *The Philosophy of the Enlightenment* (Boston: Beacon Press, 1955), p. 65.

33. J. Bronowski, *The Common Sense of Science* (Cambridge: Harvard University Press, 1955), p. 46. The mechanistic world view referred to above as scientism was commonly known in the seventeenth and eighteenth centuries as the "corpuscular philosophy," which Butterfield has described as follows: "The view became current that all the operations of nature, all the fabric of the created universe, could be reduced to the behaviour of minute particles of matter, and all the variety that presented itself to human experience could be resolved into the question of the size, the configuration, the motion, the position and the juxtaposition of these particles." Herbert Butterfield, *The Origins of Modern Science, 1300–1800* (New York: Macmillan, 1957), p. 120.

34. D'Alembert, *Éléments de Philosophie;* quoted in Cassirer, *Philosophy of the Enlightenment*, pp. 46–47. Cassirer observes that for the eighteenth century the advance of "all the intellectual sciences," involving "deeper insight into the spirit of the laws, of society, of politics, and even of poetry, seems impossible unless it is pursued in the light of the great model of the natural sciences." *Ibid.*, p. 46.

35. Carl Becker, *The Heavenly City of the Eighteenth-Century Philosophers* (New Haven: Yale University Press, paperbound ed., 1959), p. 58.

36. Becker cites some of the more extraordinary of the popularizations, such as Count Allgorotti's *Il Newtonianismo pour la dame* and Desaguliers's *The Newtonian System of the World the Best Model of Government, an Allegorical Poem. Ibid.*, p. 61.

37. Cf. Cassirer, *Philosophy of the Enlightenment*, pp. 46 ff. The degree to which "men of letters," rather more than scientists, were responsible for the corpuscular (or scientistic) philosophy has been emphasized by Butterfield, *The Origins of Modern Science*, pp. 165 ff.

38. *Entretiens sur la pluralité des mondes;* quoted in Cassirer, *Philosophy of the Enlightenment*, p. 50.

39. *Systeme de la Nature;* quoted in Randall, *Making of the Modern Mind*, p. 274. For a comprehensive analysis of Holbach's work, see Basil Willey, *The Eighteenth-Century Background* (New York: Columbia University Press, 1940), chap. IX.

40. Randall, *Making of the Modern Mind*, p. 274.

41. *L'Homme Machine;* quoted in Joseph Needham (ed.), *Science, Religion, and Reality* (New York: Braziller, 1955), p. 236. Crombie has succinctly indicated the extent to which the mechanical model came to be embraced by social scientists: "In the hands of the 'physiologists' of the French Encyclopedie like La Mettrie, D'Holbach, Condorcet and Cabanis, man became nothing but a machine; consciousness became a secretion of the brain just as bile was a secretion of the liver; and the physical and physiological laws as they conceived them were taken as the norm of the laws not only of mind but also of history and the historical progress of society. Directly descended from the Cartesian mechanical philosophy and Newtonian physics, these conceptions developed by the 18th-century French natural philosophers and sociologists became the direct ancestors of the materialist doctrines associated with Charles Darwin's theory of evolution and its sociological extensions in the 19th-century doctrine of progress." *Medieval and Early Modern Science*, pp. 313–314.

42. La Mettrie, *L'Homme Machine;* quoted in Cassirer, *Philosophy of the Enlightenment*, pp. 67, 69. Cf. the commentary on La Mettrie by the French quantum physicist Pascual Jordan, *Science and the Course of History* (New Haven: Yale University Press, 1955), pp. 108 ff.

43. Holbach, freely rendered by Cassirer, *Philosophy of the Enlightenment*, p. 69. Cf. Sir William Cecil Dampier, *A History of Science and Its Relations with Philosophy and Religion* (New York: Macmillan, rev. ed., 1943), pp. 213–214.

44. Niels Bohr, *Atomic Physics and Human Knowledge*, p. 95. Cf. Joseph Needham, "Mechanistic Biology and the Religious Consciousness," in *Science, Religion and Reality*, pp. 237 ff.

45. See Elie Halévy, *The Growth of Philosophic Radicalism* (Boston: Beacon Press, 1955), p. 8. "Are we not entitled to hold that these physiological and medical preoccupations may have helped to produce in Hartley his belief in determinism and his disposition to account scientifically for the mechanism of mental phenomena?" *Ibid.*

46. *Treatise of Human Nature* (1738), Bk. 1, Part 1, sec. IV. Cf. *Inquiry Concerning Human Understanding*, Part II, sec. 7.

47. Werner Heisenberg, *Physics and Philosophy: The Revolution in Modern Science* (New York: Harper, 1958), pp. 197–198.

48. Becker, *The Heavenly City of the Eighteenth-Century Philosophers*, p. 63.

49. "In short, nothing in life counted except what was countable:

units of weight, measure, time, space, energy, money. These were the building stones of the New World: a habitat where in the end men were acceptable only when they took on the attributes of machines, where in the foreseeable future machines would be developed to surpass and replace men." Lewis Mumford, *The Transformations of Man* (New York: Harper, 1956), p. 130. Cf. Friedrich Juenger, *The Failure of Technology* (Chicago: Regnery Gateway Editions, 1956), pp. 30 ff., *passim*.

50. Frank E. Manuel, *The New World of Henri Saint-Simon* (Cambridge: Harvard University Press, 1956), p. 119. Cf. Robert Flint, *The Philosophy of History in France and Germany* (Edinburgh: 1874), pp. 160–161.

51. "Saint Simon's apotheosis of Newton and Newton's law . . . came at the end of over a century of Newtonian faddism during which the *Principia* passed through more than twenty editions. . . . Not much mental stretching was required for a self-taught latter-day *philosophe* like Saint-Simon, brought up in the crude materialist school of the eighteenth century, to make the law of gravity cover moral and spiritual phenomena as well. The sensationalists had shown that the real origin of ideas was in matter, and matter obeyed Newton's law; why then should ideas be exempt from its dominion?" Manuel, *The New World of Henri Saint-Simon*, p. 119.

52. *Oeuvres de Saint-Simon et d'Enfantin*; quoted in F. A. Hayek, *The Counter-Revolution of Science: Studies in the Abuse of Reason* (Glencoe, Ill.: Free Press, 1952), p. 120.

53. *Ibid.*, p. 121.

54. *Ibid.*, pp. 122, 126.

55. Harriet Martineau, trans., *The Positive Philosophy of Auguste Comte* (London, 1893), Vol. II, p. 61. Cf. Robert E. Park and Ernest W. Burgess, *Introduction to the Science of Sociology* (Chicago: University of Chicago Press, 1921), pp. 3–4.

56. "A new lust for domination over man, shaped after the pattern of domination over nature, had developed in technique-minded men. The Saint-Simonists set forth, ultimately, a construct in which men are controlled with a precision reminiscent of the engineer's methods. The highly emotional humanitarianism which pervades the system did not blind everybody to the fact that a new imperialism, a new lust for absolute power was finding expression." Yves R. Simon, *Philosophy of Democratic Government* (Chicago: University of Chicago Press, 1951), p. 292. Cf. Albert Salomon, *The Tyranny of Progress: Reflections on the Origins of Sociology* (New York: Noonday Press, 1955), pp. 100 ff. See also, on the Saint-Simonians generally, Sheldon Wolin, *Politics and Vision* (Boston: Little, Brown, 1960), pp. 376 ff.

57. As Mumford has noted, "The people who are honored as

benefactors of the race in [Bacon's] fragmentary utopia, The New Atlantis, are all scientists or inventors of the practical arts. . . . Feelings, emotions, states of mind, were becoming unreal to him and his scientifically minded contemporaries; they faced the perils of subjective hallucination by closing up that side of the personality and declaring it utterly bankrupt." *The Condition of Man*, p. 243. Cf. Mumford, *The Story of Utopias* (New York, 1922).

58. Halévy, *The Rise of Philosophic Radicalism*, p. 6. Halévy opens his history of the movement with these words: "On the one hand the development of the physical sciences, the discovery of Newton's principle which made it possible to found on a single law a complete science of nature, and the conception of the hope of discovering an analogous principle capable of serving for the establishment of a synthetic science of the phenomena of moral and social life; on the other hand a profound crisis in society . . . these are general causes of the formation of Philosophical Radicalism." *Ibid.*, p. 3. Cf. Leslie Stephen, *The English Utilitarians* (London: Duckworth, 1900), Vol. II, p. 311, *passim*. For further treatment of the Utilitarian psychology, see below, Chap. II.

59. Elliott and MacDonald, *Western Political Heritage*, p. 704.

60. Cf. Tawney, *Religion and the Rise of Capitalism*, p. 160.

61. Karl Polanyi, *The Great Transformation* (New York: Rinehart, 1944), chap. 10. Polanyi's title, of course, embraces a second transformation as well: the breakdown of the market system and its philosophical framework in the twentieth century. (I am indebted to Ashley Montagu for correction of Townsend's name, erroneously cited by Polanyi as *William* Townsend.)

62. *Ibid.*, pp. 115–116. "As gradually the laws governing a market economy were apprehended, these laws were put under the authority of Nature herself. The law of diminishing returns was a law of plant physiology. The Malthusian law of population reflected the relationship between the fertility of man and that of the soil. . . . The laws of a competitive society were put under the sanction of the jungle." *Ibid.*, p. 125.

63. *Ibid.*, pp. 125–126.

64. Edgeworth, *Mathematical Psychics* (London, 1881). On Edgeworth, see John Maynard Keynes, *Essays in Biography* (New York: Horizon Press, 1951).

65. For an informative expansion on this theme, see Robert L. Heilbroner, *The Worldly Philosophers* (New York: Simon and Schuster, 1953), chap. VII.

66. Cf. Halévy, *The Rise of Philosophic Radicalism*, pp. 489 ff., where the paradox is described as a conflict between the "Westminster philosophy of the artificial identification of interests" and the "Manchester

philosophy of the natural identification of interests." In general, the Manchesterites prevailed.

67. Lord Lindsay has observed that the same mechanistic assumption of a natural and automatic harmony of interests (to be achieved once the monkey wrench has been discovered and thrown out) underlay such diverse theories of the nineteenth century as Benthamism and Marxism, which go back to Hobbes in treating politics as a branch of physics with individuals as its hapless atoms. A. D. Lindsay, *The Modern Democratic State* (Oxford, 1943), Vol. I, pp. 82 ff.

68. *Darwin, Marx, Wagner* (Boston: Little, Brown, 1941), p. 69. A modern biologist adds: "The theory of evolution gave rise to the hope that it would be possible to explain on mechanical lines not merely the origin and tansformation of living species, but also the genesis of psychic life and human society, eliminating the intervention of supernatural causes." Needham, *Science, Religion and Reality*, p. 159.

69. "On the Physical Basis of Life," in Huxley, *Methods and Results: Essays* (London, 1893), pp. 130–165. Cf. Huxley, *Evolution and Ethics and Other Essays* (New York: Humboldt, 1894). For an acid discussion of Huxley's mechanical philosophy, see L. Susan Stebbing, *Philosophy and the Physicists* (New York: Dover, 1958), pp. 145 ff. A much more sympathetic account is given by William Irvine, *Apes, Angels and Victorians: A Joint Biography of Darwin and Huxley* (London: Readers Union, 1956).

70. Herbert Spencer, *First Principles* (New York: Appleton, 1864), pp. 284, 407. It took William James to translate Spencer's ponderous proposition into American: "Evolution is a change from a no-howish, untalkaboutable, all-alikeness to a somehowish and in general talkaboutable not-all-alikeness by continuous stick-togetherations and somethingelsifications." Quoted in Gilbert Highet, *The Art of Teaching* (New York: Knopf, 1954), p. 207.

71. *First Principles*, pp. 340–371; cited in Richard Hofstadter, *Social Darwinism in American Thought* (Boston: Beacon Press, 1955), p. 37.

72. Spencer, *Social Statics* (New York: Appleton, 1864 [orig. ed. 1850]), pp. 79–80.

73. William Graham Sumner and Albert G. Keller, *The Science of Society* (New Haven: Yale University Press, 1927), Vol. I, pp. 40–41.

74. "The Absurd Effort to Make the World Over," in *Essays of William Graham Sumner*, Albert G. Keller and Maurice R. Davie, eds. (New Haven: Yale University Press, 1934), Vol. I, p. 105.

75. *What Social Classes Owe to Each Other* (New York: Harper, 1883), p. 135.

76. *Essays of William Graham Sumner*, p. 162; *Folkways* (Boston: Ginn, 1907), p. 201. Cf. Emory S. Bogardus, *The Development of Social Thought* (New York: Longmans, Green, 1940), p. 334.

77. Sumner, "The Mores of the Present and the Future," *Yale Review*, Vol. XVIII (1909–10), pp. 235–236; *What Social Classes Owe to Each Other*, p. 25; *Folkways*, p. 65.

78. "The eighteenth-century notions about equality, natural rights, classes and the like produced nineteenth-century states and legislation, all strongly humanitarian in faith and temper; at the present time the eighteenth-century notions are disappearing, and the *mores* of the twentieth century will not be tinged by humanitarianism as those of the last hundred years have been." *Essays of William Graham Sumner*, pp. 65–67.

79. Cited in Philip F. Wiener, *Evolution and the Founders of Pragmatism* (Cambridge: Harvard University Press, 1949), pp. 35, 57.

80. See discussion in Dampier, *A History of Science*, p. 342. Haeckel's major work is *The Riddle of the Universe* (New York: Harper, 1900).

81. Ludwig Gumplowicz, *The Outlines of Sociology* (Philadelphia: American Academy of Political and Social Science, 1899), p. 156.

82. *Ibid.*, pp. 148, 156–157.

83. *Ibid.*, p. 213. "Hence the alpha and omega of sociology, its highest perception and final word, is: human history a natural process. . . . By contributing to the knowledge of these laws sociology lays the foundation for the morals of reasonable resignation, higher than those resting on imaginary freedom and self-determination and resulting in the inordinate overestimation of the individual and those unreasonable aspirations which find expression in horrible crimes against the natural law of order." *Ibid.*

84. *Ibid.*, p. 190.

85. *Ibid.*, p. 180.

86. *Ibid.*

87. Ernst Cassirer, *An Essay on Man* (New York: Doubleday Anchor Books, 1953), p. 38.

88. Cf. Popper, *The Open Society and Its Enemies*, pp. 83–84. On the mechanical formula of Marxist "scientism," see also Morgenthau, *Dilemmas of Politics*, p. 242.

89. The quoted passage is from the *Anti-Duhring* (1877). Elsewhere, in a famous statement, Engels wrote: "Thus dialectics reduced itself to the science of the general laws of motion—both of the external world and of human thought—two sets of laws which are identical in substance, . . . Thereby the dialectic of the concept itself became merely the conscious reflex of the dialectical motion of the real world and the dialectic of

Hegel was placed upon its head; or rather, turned off its head, on which it was standing before, and placed upon its feet again." V. Adoratsky (ed.), *Karl Marx: Selected Works* (New York: International Publishers, n.d.), Vol. I, pp. 454, 455. Cf. *ibid.*, Vol. I, pp. 25, 27, 157, 162, 436, 455 ff.

90. "But if the . . . question is raised: what then are thought and consciousness, and whence they come, it becomes apparent that they are products of the human brain, and that man himself is a product of nature, which has been developed in and along with its environment; whence it is self-evident that the products of the human brain, being in the last analysis also products of nature, do not contradict the rest of nature, but are in correspondence with it." Engels, *Anti-Duhring*; in Adoratsky (ed.), *ibid.*, p. 25. See also Engels, *Dialectics of Nature* (New York: International Publishers, 1940), p. 35.

91. For discussion of this movement and the subsequent reaction against it, see Floyd W. Matson, "History as Art: The Psychological-Romantic View," *Journal of the History of Ideas*, Vol. XVIII, No. 2 (April, 1957), pp. 270–279.

92. J. B. Bury, "The Science of History," in *Selected Essays* (London, 1930), p. 3. Well before that, Taine had claimed for his study of the *ancien régime:* "The historian may be permitted the privilege of the naturalist; I have observed my subject as one might observe the metamorphosis of an insect." And Fustel de Coulanges, asserting that the mathematics of Descartes had converted history into an objective science, reproved his audience: "Do not applaud me. It is not I who address you, but history that speaks through my mouth." Emery Neff, *The Poetry of History* (New York: Columbia University Press, 1947), p. 192.

93. Henry Adams, "The Tendency of History," in *The Degradation of the Democratic Dogma* (New York, 1920), p. 126. Cf. *The Education of Henry Adams* (Boston, 1918), pp. 382, 431, *passim*. On Adams' historical scientism, see Henry S. Kariel, "The Limits of Social Science: Henry Adams' Quest for Order," *American Political Science Review*, Vol. I, No. 4 (December, 1956), pp. 1074–1192.

94. *The Origins of Psychoanalysis: Sigmund Freud's Letters* (New York: Basic Books, 1954), p. 355.

95. For a more detailed assessment of Freud, see Chap. VI, below.

96. Ludwig von Bertalanffy, *Problems of Life* (New York: Harper Torchbooks, 1960), p. 202.

CHAPTER II *The Alienated Machine*

1. Thus David Easton, referring to students of political behavior, states that "we can call the group behavioralists, so long as we do not confuse their views with that of strict, and now antiquated, behaviorism." *The Political System: An Inquiry into the State of Political Science* (New York: Knopf, 1953), p. 151. See also *ibid.*, p. 202. Cf. Easton's contribution to the symposium, *The Limits of Behavioralism in Political Science* (Philadelphia: American Academy of Political and Social Science, October, 1962), p. 3.

2. One clue to this is provided by the fact that much, if not most, current literature on the "science of behavior" is contributed by psychologists in the behaviorist tradition. Most notable perhaps are the works of Clark L. Hull and B. F. Skinner, discussed below. Cf. Charles Osgood, "Behavior Theory and the Social Sciences," *Journal of Behavioral Science* (July, 1956), pp. 168 ff.

3. Quoted in John B. Watson, *Behaviorism* (Chicago: University of Chicago Press, 1958), p. iii.

4. Kenneth W. Spence, in Herbert Feigl and May Brodbeck (eds.), *Readings in the Philosophy of Science* (New York: Appleton-Century-Crofts, 1953), p. 571. See also Gustav Bergmann, "The Contribution of John B. Watson," in Jordan Scher (ed.), *Theories of the Mind* (New York: Free Press, 1962), pp. 674 ff.

5. Edna Heidbreder, *Seven Psychologies* (New York: Appleton-Century, 1935), p. 40. "With the mechanical viewpoint, the notion of bodies as bits of matter moving in space and time, Hobbes built up the scheme of human nature as a purely mechanical thing, avoiding altogether the interactionism of Descartes. It is no exaggeration to say that Hobbes took the whole fabric of the seventeenth-century physical view of the world and fashioned from it a conception of human nature. Every thought, feeling, and purpose was simply internal *motion*." Gardner Murphy, *Historical Introduction to Modern Psychology*, rev. ed. (New York: Harcourt, Brace, 1951), pp. 24–25. Cf. G. C. Robertson, "Hobbes," *Encyclopedia Britannica*, 11th ed., Vol. XIII, p. 552.

6. Locke's position is enunciated in his *Essay Concerning Human Understanding* (1690). Cf. H. C. Warren, *A History of the Association Psychology* (New York: Scribner's, 1921), pp. 36–40; Edwin G. Boring, *A History of Experimental Psychology* (New York: Appleton-Century-Crofts, 2nd ed., 1950), pp. 168–176. Allport summarizes his contemporary relevance as follows: "The Lockean point of view . . . has been and is still dominant in Anglo-American psychology. Its representatives are found in associationism of all types, including environmentalism, behaviorism, stimulus-response (familiarly abbreviated as S-R) psychology, and all other

stimulus-oriented psychologies, in animal and genetic psychology, in positivism and operationism, in mathematical models—in short, in most of what today is cherished in our laboratories as truly 'scientific' psychology." Gordon W. Allport, *Becoming: Basic Considerations for a Psychology of Personality* (New Haven: Yale University Press, 1955), pp. 8–9.

7. Isaiah Berlin, *The Age of Enlightenment* (New York: Mentor Books, 1956), p. 18.

8. Warren, *A History of the Association Psychology*, pp. 87–88.

9. See Chapter 1, above.

10. Elie Halévy, *The Growth of Philosophic Radicalism* (Boston: Beacon Press, 1955), p. 8.

11. Heidbreder, *Seven Psychologies*, p. 62. Cf. Boring, *A History of Experimental Psychology*, pp. 214–216.

12. *The Growth of Philosophic Radicalism*, p. 439. Cf. Leslie Stephen, *The English Utilitarians* (London: Duckworth, 1900), Vol. II, p. 311, *passim.*

13. Halévy, *The Growth of Philosophic Radicalism*, p. 440.

14. Warren, *A History of the Association Psychology*, p. 163.

15. Halévy, *The Growth of Philosophic Radicalism*, p. 458. Cf. Stephen, *The English Utilitarians*, p. 290: "The pith of Mill's book [*Analysis of the Phenomena of the Human Mind*] is thus determined. His aim is to give a complete analysis of mental phenomena, and therefore to resolve those phenomena into their primitive constituent atoms."

16. An excellent source is still Harriet Martineau's abridged translation, *The Positive Philosophy of Auguste Comte* (London, 1893). For pertinent commentaries, in addition to works cited below, see Nicholas S. Timasheff, *Sociological Theory: Its Nature and Growth* (Garden City, N. Y.: Doubleday, 1955), pp. 15–30; Edward Caird, *The Social Philosophy and Religion of Comte* (Glasgow, 1893); John H. Hallowell, *Main Currents in Modern Political Thought* (New York: Henry Holt, 1950), chap. 9.

17. Quoted in F. A. Hayek, *The Counter-Revolution of Science* (Glencoe, Ill.: Free Press, 1952), p. 172. Hayek's study is in large part an examination of the social and political implications of Comte and the Saint-Simonian school generally.

18. *Ibid.*

19. *Ibid.*

20. Professor Gordon W. Allport has made me aware that the later Comte did (if obliquely) smuggle the person back into psychology by virtue of his concept of "morale." The point is dealt with in Allport, "The Historical Background of Modern Social Psychology," in G. Lindzey (ed.),

Handbook of Social Psychology (Cambridge: Addison-Wesley, 1954), Vol. I, chap. 1.

21. "Many a scientist has patiently designed experiments for the *purpose* of substantiating his belief that animal operations are motivated by no purposes. He has perhaps spent his spare time in writing articles to prove that human beings are as other animals so that 'purpose' is a category irrelevant for the explanation of their bodily activities, his own activities included. Scientists animated by the purpose of proving that they are purposeless constitute an interesting subject for study." Alfred North Whitehead, quoted in J. W. N. Sullivan, *The Limitations of Science* (New York: Mentor Books, 1949), p. 126.

22. Albert Salomon, *The Tyranny of Progress: Reflections on the Origins of Sociology* (New York: Noonday Press, 1955), p. 101.

23. Quoted in Salomon, *ibid.*, p. 100.

24. "It is the exclusive property of the positive principle to recognize the fundamental law of continuous human development, representing the existing evolution as the necessary result of the gradual series of former transformations, by simply extending to social phenomena the spirit which governs the treatment of all other natural phenomena." *Social Physics: From the Positive Philosophy of Auguste Comte,* Harriett Martineau, trans. (New York: Blanchard, 1850), p. 432.

25. *Ibid.*, p. 435.

26. Salomon, *The Tyranny of Progress*, p. 103.

27. *Ibid.*, p. 104.

28. On Comte's attempt to found a religion of Humanity, see Hallowell, *Main Currents in Modern Political Thought*, chap. 9.

29. Cf. Heidbreder, *Seven Psychologies*, pp. 237–238.

30. "And fundamentally it is the attempt to study human animals as other animals are studied that constitutes the true originality of behaviorism. . . . [This] means that man must be regarded as an animal species, simply as one among the many species of the animal order, in no fundamental sense constituting a special case." *Ibid.*, p. 238.

31. *Historical Introduction to Modern Psychology*, p. 253.

32. *Ibid.*

33. The conditioned reflex is sufficiently familiar to require no explanation here. See I. P. Pavlov, *Conditioned Reflexes: An Investigation of the Physiological Activity of the Cerebral Cortex*, G. V. Anrep, trans. (New York: Oxford University Press, 1927). For history and commentary, cf. Boring, *A History of Experimental Psychology*, pp. 635 ff.; B. F. Skinner, *Science and Human Behavior* (New York: Macmillan, 1953), chap. IV; Murphy, *Historical Introduction to Modern Psychology*, pp. 255 ff. A

sharply critical account is given by Robert E. L. Faris, *Social Psychology* (New York: Ronald Press, 1952), pp. 77 ff.

34. John Dewey, "The Reflex Arc Concept in Psychology," *Psychological Review* (1896), Vol. III, pp. 357–370.

35. Pavlov "rejected all appeals regarding the formulation of psychological problems and warned his students to keep away from psychology. Though he had become within a decade a hero of the rapidly developing psychology of objective-behavior study, his response to an invitation to attend the International Congress of Psychology in 1929 was that he doubted whether psychologists would really be interested in what he had to say." Murphy, *Historical Introduction to Modern Psychology*, p. 257. Cf. I. P. Pavlov, "The Reply of a Physiologist to Psychologists," *Psychological Review*, Vol. XXXIX (1932), pp. 91–127.

36. His doctoral dissertation bore the title "Kinaesthetic and Organic Sensations: Their Role in the Reactions of the White Rat to the Maze," *Psychological Review* (1907), Vol. VIII, Monograph Supplement No. 2, pp. 1–100.

37. See John B. Watson, *Behaviorism* (Chicago: University of Chicago Press, 1958 [orig. ed. 1924]), p. 5.

38. John B. Watson, *Behavior: An Introduction to Comparative Psychology* (New York: Henry Holt, 1914), p. 27.

39. *Seven Psychologies*, p. 235.

40. Watson, *Behaviorism*, pp. 5–6.

41. *Ibid.*, p. 6.

42. *Ibid.*

43. "By stimulus we mean any object in the general environment or any change in the tissues themselves . . . such as the change we get when we keep an animal from sex activity, when we keep it from feeding, when we keep it from building a nest. By response we mean anything the animal does—such as turning toward or away from a light, jumping at a sound, and more highly organized activities such as building a skyscraper, drawing plans, having babies, writing books, and the like." *Ibid.*

44. *Ibid.*, p. 8.

45. *Ibid.*, p. 11.

46. *Ibid.*, p. 269.

47. John B. Watson, *Psychology from the Standpoint of a Behaviorist* (Philadelphia: Lippincott, 3rd. ed., 1929), p. 7.

48. John B. Watson, *The Ways of Behaviorism* (New York: Harper, 1926), pp. 9, 35. (See headnote to this chapter.) Cf. Watson, *Behaviorism*, p. 104.

49. For details, see John K. Winkler and Walter Bromberg, *Mind Explorers* (New York: Reynal and Hitchcock, 1939), pp. 300 ff.; A. A. Roback, *Behaviorism at Twenty-Five* (Cambridge: Sci-Art Publishers, 1937), p. 175.

50. Watson, *Behaviorism*, p. 44.

51. *Ibid.*, p. 303.

52. *Ibid.*

53. *Ibid.*

54. *Seven Psychologies*, p. 257.

55. Friedrich Georg Juenger, *The Failure of Technology* (Chicago: Gateway Editions, 1956), p. 142.

56. Quoted in Robert S. Woodworth, *Contemporary Schools of Psychology* (New York: Ronald Press, rev. ed., 1948), pp. 92–93. Woodworth adds: "Behaviorism meant to many young men and women of the time a new orientation and a new hope when the old guides had become hopelessly discredited in their eyes. It was a religion to take the place of religion." *Ibid.*, p. 94.

57. "In the case of man, all healthy individuals, as we saw in our discussion of Instincts, start out *equal*. Quite similar words appear in our far-famed Declaration of Independence. The signers of that document were nearer right than one might expect, considering their dense ignorance of psychology." Watson, *Behaviorism*, p. 270.

58. Karl Mannheim, *Man and Society in an Age of Reconstruction* (New York: Harcourt, Brace, 1950), pp. 51 ff.

59. *The Failure of Technology*, pp. 141, 142. Writing of the trend toward organized procedures in all aspects of life, Roderick Seidenberg declares prophetically: "Under the momentum of this universal trend, the individual will indeed find himself churned into an ever smaller particle, into a minute and at length irreducible atom of the social system. As the significance of the individual is thus steadily diminished, his status and identity must necessarily approach that of a statistical average. . . ." *Post-Historic Man: An Inquiry* (Boston: Beacon Press, paperback edition, 1957), p. 13.

60. John MacMurray, *The Boundaries of Science* (London: Faber and Faber, 1939), pp. 125, 127.

61. *Man and Society in an Age of Reconstruction*, pp. 214–216. The psychologist William MacDougall, a bitter antagonist of Watson during the twenties, put the ultimate case against behaviorism in moral terms: "But I do suggest and contend that the crude materialistic theory of human nature, the theory that man is a machine and nothing more, . . . cannot fail in the long run to contribute very considerably to the decay of morals and

the increase of crime. For it is a theory utterly incompatible with any view of man as a responsible moral being . . . ; a theory which represents man as incapable of choosing between good and evil, as the purely passive sport of circumstances over which he has no control; a theory which, if it is accepted, must make all talk of self-control, of self-improvement, or purposes and ideals seem sheer nonsense, survivals from an age of naive ignorance." "The Psychology They Teach in New York," in William P. King (ed.), *Behaviorism: A Battle Line* (Nashville: Cokesbury Press, 1930), pp. 33–34.

62. Hannah Arendt, *The Human Condition* (Chicago: University of Chicago Press, 1958), p. 45. "If economics is the science of society in its early stages, when it could impose its rules of behavior only on sections of the population and on parts of their activities, the rise of the 'behavioral sciences' indicates clearly the final stage of this development, when mass society has devoured all strata of the nation and 'social behavior' has become the standard for all regions of life." *Ibid.*

63. Boring, *A History of Experimental Psychology*, p. 645.

64. Murphy, *Historical Introduction to Modern Psychology*, p. 267.

65. Woodworth, *Contemporary Schools of Psychology*, p. 95.

66. Quoted in Woodworth, *ibid.*, p. 215.

67. Perhaps the best-known of these defectors is Karl S. Lashley, whose early views were summed up in the statement that "the essence of behaviorism is the belief that the study of man will reveal nothing except what is adequately describable in the concepts of mechanics and chemistry." Lashley, "The Behavioristic Interpretation of Consciousness," *Psychological Review*, Vol. 30 (1923), pp. 237–272, 329–353. By 1931 Lashley had abandoned what he came to call "muscle-twitch psychology" and was seriously questioning conventional notions of S-R and conditioned reflexes. See his article, "Cerebral Control versus Reflexology," *Journal of General Psychology*, Vol. V (1931), pp. 3–20.

68. See George H. Mead, *Mind, Self and Society* (Chicago: University of Chicago Press, 1934), pp. 10–11, 100–109.

69. See the recent collection of his articles: Edward Chace Tolman, *Behavior and Psychological Man* (Berkeley: University of California Press, Paperbound Edition, 1958), esp. "Foreword," pp. vii, ix.

70. Concerning the Yale Institute of Human Relations, Hall and Lindzey write: "This institution was established in 1933 under the direction of Mark May in an effort to bring about closer collaboration and integration among psychology, psychiatry, sociology, and anthropology. . . . The first decade of its existence represents one of the most fruitful periods of collaboration in the behavioral sciences that has occurred in any American university. Although Clark Hull provided the theoretical underpinning for

this group, its activities were by no means focused primarily on experimental psychology. . . . Among the outstanding members of this group are John Dollard, Ernest Hilgard, Carl Hovland, Donald Marquis, Neal Miller, O. H. Mowrer, Robert Sears, Kenneth Spence, and John Whiting." Calvin S. Hall and Gardner Lindzey, *Theories of Personality* (New York: Wiley, 1957), p. 424.

71. Clark L. Hull, *Principles of Behavior: An Introduction to Behavior Theory* (New York: Appleton-Century, 1943), p. v. (Emphasis added.)

72. *Ibid.*, p. 24.

73. *Ibid.*, pp. 24–25. Cf. Hull's later classification of eight different "major automatic adaptive behavior mechanisms." Clark L. Hull, *A Behavior System: An Introduction to Behavior Theory Concerning the Individual Organism* (New Haven: Yale University Press, 1952), pp. 348 ff.

74. *Principles of Behavior*, pp. 25–26. See also *ibid.*, p. 29.

75. *Ibid.*, p. 26.

76. *Ibid.*, p. 27. (Emphasis added.)

77. *Ibid.*, pp. 27–28.

78. *Ibid.*, p. 28.

79. *Ibid.*, p. 32.

80. *Ibid.*, p. 398. (Emphasis added.) Cf. Clark L. Hull *et al.*, *Mathematico-Deductive Theory of Rote Learning: A Study in Scientific Methodology* (New Haven: Yale University Press, 1940).

81. *Principles of Behavior*, p. 399.

82. *Ibid.*

83. *Ibid.*, p. 400. (Emphasis added.)

84. Discussing this "scientistic" tendency in the wider context of the social sciences, Thomas I. Cook has written: "Innumerable observations are made, endless measurements taken, often without clear purpose or subsequent assessment of their place in a whole body of knowledge and of what they actually prove. . . . Yet the underlying faith, which such results are held to sustain and warrant, is that ultimately the whole of man's behavior, individual and collective, can be thus descriptively interpreted and predictive laws therefor and thereby attained." "The Methods of Political Science, Chiefly in the United States," in *Contemporary Political Science: A Survey of Methods, Research, and Teaching* (Paris: UNESCO, 1950), pp. 75–76.

85. "There will be encountered vituperative opposition from those who cannot or will not think in terms of mathematics, from those who prefer to have their scientific pictures artistically out of focus, from those

who are apprehensive of the ultimate exposure of certain personally cherished superstitions and magical practices, and from those who are associated with institutions whose vested interests may be fancied as endangered." *Principles of Behavior*, p. 401.

86. *Ibid.*, p. 400.

87. *Ibid.*

88. *Ibid.*, pp. 400–401.

89. *Ibid.*, p. 401. (Emphasis added.)

90. In addition to the discussion of Comte in this chapter, see especially Chap. 1, "Social Scientism in the Century of Progress," pp. 31 ff.

91. *A Behavior System*, p. 332.

92. *Ibid.*, p. 339.

93. *Ibid.*, p. 340.

94. *Ibid.*, p. 345.

95. For a similar critique of the robot psychology of Hull and others, cf. Michael Polanyi, *Personal Knowledge*, pp. 263–264, 369–373.

96. Compare the statement of S. S. Stevens, a ranking experimental psychologist: "What we can get at in the study of living things are the responses of organisms, not some hyperphysical mental stuff, which, by definition, eludes objective test." "Measurement and Man," *Science*, Vol. CXXVII (1958), p. 386. For discussion of the concept of "self" in psychology, cf. Allport, *Becoming*, pp. 36 ff.

97. See Woodworth, *Contemporary Schools of Psychology*, p. 116; Boring, *A History of Experimental Psychology*, pp. 650 ff.

98. B. F. Skinner, *The Behavior of Organisms: An Experimental Analysis* (New York: Appleton-Century, 1938), p. 5.

99. *Ibid.*

100. *Ibid.*, p. 8.

101. *Ibid.*, chap. 13. See also the discussion of his later works below.

102. *Ibid.*, p. 47. Cf. the "cozy rodent" description of Tolman, "A Stimulus-Expectancy Need-Cathexis Psychology," *Science*, Vol. CI (1945), p. 166.

103. *The Behavior of Organisms*, p. 47.

104. *Ibid.*, p. 442. An observation of the late Kurt Riezler sheds light on the behaviorists' predilection for rodents: "A scientist who, aiming at objectivity whatever the cost, excludes not merely his own subjectivity but also the subjective factors of his subject matter, deprives himself of the greater part of his facts, and his subject matter of its specific nature.

He puts a ban on any kind of interpretation, restricts himself to registering 'overt actions,' and finally prefers the behavior of rats to that of a human soul, since the latter but not the former might suggest an interpretation that could be deemed subjective." Kurt Riezler, "Some Critical Remarks on Man's Science of Man," *Social Research* (November, 1945), p. 485.

105. Paul LaFitte, *The Person in Psychology: Reality or Abstraction* (London: Routledge and Kegan Paul, 1957), pp. 31–32.

106. *Ibid.*, p. 32.

107. Thus Gordon W. Allport observes: "While we righteously scorn what one of us has called the 'subjective, anthropomorphic hocus-pocus of mentalism,' we would consider a colleague emotional and mystical should he dare to speak of 'the objective mechanomorphic hocus-pocus of physicalism.'" Gordon W. Allport, *The Nature of Personality: Selected Papers* (Cambridge: Addison-Wesley Press, 1950), p. 187.

108. Joseph Wood Krutch, *The Measure of Man* (New York: Grosset and Dunlap, 1953), p. 106.

109. B. F. Skinner, *Science and Human Behavior* (New York: Macmillan, 1953), p. 5.

110. *Ibid.*, pp. 5–6. (Emphasis added.) Cf. the remark of the central character in Skinner's *Walden Two* (New York: Macmillan, 1948), p. 214: "I deny that freedom exists at all. I must deny it—or my program would be absurd. You can't have a science about a subject matter which hops capriciously about."

111. *Science and Human Behavior*, p. 6.

112. *Ibid.*, p. 9. (Emphasis added.)

113. *Ibid.*, pp. 10–11.

114. *Ibid.*, p. 29. "The fictional nature of this form of inner cause is shown by the ease with which the mental process is discovered to have just the properties needed to account for the behavior. . . . In all this it is obvious that the mind and the ideas, together with their special characteristics, are being invented on the spot to provide spurious explanations. A science of behavior can hope to gain very little from so cavalier a practice." *Ibid.*, p. 30.

115. *Ibid.*, p. 31.

116. *Ibid.*, p. 35.

117. *Ibid.*, pp. 35–36.

118. *Ibid.*, p. 37. See the illuminating, as well as entertaining, criticism of such laboratory objectivism by Anthony Standen, *Science Is a Sacred Cow* (New York: Dutton, 1958), pp. 120–121. Skinner is not unique in his eagerness to set about the experimental manipulation of human sub-

jects. The following statement of a "behavioral criminologist," Hans von Hentig, is representative: "Is criminology a science? If we mean by *science* a body of learnable and teachable knowledge that can be applied with a reasonable degree of certainty to modifying life, criminology is on the way to becoming a science. Physicists, physiologists, and others will deny this contention and point out that our material cannot be manipulated in test tubes and cannot be carried through experiments and counterexperiments. The objection is not just. We could very well manipulate our human material in test situations if we dared to do it. Experimental conditions could be devised and turned to good account. It is true that our experimental schemes cannot match the arrangements of the physicist, for social forces are not as easily isolated, set out, and reshuffled as are physical elements. Yet it would not be difficult to develop a technique of social experimentation. . . ." Hans von Hentig, *Crime: Causes and Conditions* (New York: McGraw-Hill, 1947), p. 1.

119. *Science and Human Behavior*, p. 38.

120. *Ibid.*, p. 42.

121. *Ibid.*, p. 315. The unblinking objectivism of Skinner's anatomy of terror and force as techniques of behavioral control would be awesome if it were not faintly ridiculous, rather like an updated parody of Machiavelli. For example: "The use of force has obvious disadvantages as a controlling technique. It usually requires the sustained attention of the controller. . . . It generates strong emotional dispositions to counterattack. It cannot be applied to all forms of behavior; handcuffs restrain part of a man's rage but not all of it. . . . For all these reasons, control through physical restraint is not so promising a possibility as it may at first appear." *Ibid.*, pp. 315–316.

122. *Ibid.*, p. 322. (Emphasis added.)

123. *Ibid.*, pp. 430 ff. Cf. Skinner, "The Control of Human Behavior," *Transactions of the New York Academy of Sciences*, Ser. II, Vol. XVII (May, 1955): "Eventually, the practices which make for the greatest biological and psychological strength of the group will presumably survive, as will the group which adopts them. Survival is not a criterion which we are free to accept or reject [sic] but it is, nevertheless, the one according to which our current decisions will eventually be tested. It is less clear-cut than some absolute criterion of right and wrong, but it is more reassuring in its recognition of the changing needs of society."

124. *Science and Human Behavior*, p. 435.

125. *Ibid.*, p. 436.

126. *Ibid.*, p. 437.

127. *Ibid.*, p. 438. (Emphasis added.) "The issue," Skinner has written elsewhere, "is not between freedom and control but between one sort of

control and another." "The Concept of Freedom from the Point of View of a Science of Human Behavior," unpublished paper for private distribution (1955, mimeo.), p. 6.

128. *Science and Human Behavior*, p. 439.

129. *Ibid.*, p. 148.

130. See *Ibid.*, chap. 18, "The Self," pp. 283 ff.

131. *Ibid.*, p. 443.

132. "The Control of Human Behavior," p. 8. Cf. the protagonist's laconic comment in *Walden Two*, p. 220: "The despot must wield his power for the good of others. If he takes any step which reduces the sum total of human happiness, his power is reduced by a like amount. What better check against a malevolent despotism could you ask for?"

133. *Ibid.*, p. 444.

134. *Ibid.*, p. 445.

135. See Skinner's discussion, *ibid.*, p. 445. An especially effective and authoritative demonstration of the scientific flaw, as well as the ethical implications, of all theories which equate *survival* with *good*—whether the survival be that of the individual or of the group—is to be found in George Gaylord Simpson, *The Meaning of Evolution* (New York: Mentor Books, 1951), pp. 148 ff. Cf. below, Chap. V, "The Biology of Freedom."

136. *Science and Human Behavior*, pp. 447–448. (Emphasis added.)

137. *Ibid.*, p. 449. Skinner concludes his article on "The Control of Human Behavior" (cited above) with a similar warning against "wasting time" on talk about freedom and individual responsibility: "Unless there is some unseen virtue in ignorance, our growing understanding of human behavior will make it all the more feasible to design a world adequate to the needs of men. But we cannot gain this advantage if we are to waste time defending outworn conceptions of human nature, conceptions which have long since served their original purpose of justifying special philosophies of government. A rejection of science at this time, in a desperate attempt to preserve a loved but inaccurate conception of man, would represent an unworthy retreat in man's continuing effort to build a better world."

138. Krutch, *The Measure of Man*, *passim*. Cf. Glenn Negley and J. Max Patrick, *The Quest for Utopia* (New York: Schuman, 1952), pp. 589–590.

139. E.g.: "I wish I could convince you of the simplicity and adequacy of the experimental point of view. . . . What is the 'original nature of man?' . . . That's certainly an experimental question—for a science of behavior to answer. And what are the techniques, the engineering practices, which will shape the behavior of the members of a group so that they will

function smoothly for the benefit of all? That's also an experimental question, Mr. Castle—to be answered by a behavioral technology. . . . Experimentation with life—could anything be more fascinating?" *Walden Two,* p. 145.

140. *Ibid.,* p. 43.

141. *Ibid.,* p. 260. Something of what that cultural revision would entail is suggested by the following comment of the founder of Walden Two: "In Walden Two no one worries about the government except the few to whom that worry has been assigned. To suggest that everyone should take an interest would seem as fantastic as to suggest that everyone should become familiar with our Diesel engines. Even the constitutional rights of the members are seldom thought about, I'm sure. The only thing that matters is one's day-to-day happiness and a secure future." *Ibid.,* p. 225.

CHAPTER III   *The Manipulated Society*

1. Karl Mannheim, *Ideology and Utopia: An Introduction to the Sociology of Knowledge* (New York: Harcourt, Brace, 1959), p. 39. Cf. Mannheim's earlier discussion, published as a book review in the *American Journal of Sociology,* Vol. XXXVIII, No. 2 (September, 1932), pp. 273–282; reprinted in Mannheim, *Essays on Sociology and Social Psychology* (New York: Oxford University Press, 1953), pp. 185–195.

2. See, for example, his *Studies in the Theory of Human Society* (New York: Macmillan, 1922), esp. chap. XV. "Pluralistic behavior is the subject-matter of the psychology of society, otherwise called sociology, a science statistical in method, which attempts, first, to factorize pluralistic behavior, and second, to explain its genesis, integration, differentiation, and functioning by accounting for them in terms of the variables (1) stimulation, and (2) the resemblance (more or less) to one another of reacting mechanisms." *Ibid.,* p. 252.

3. Nicholas S. Timasheff, *Sociological Theory: Its Nature and Growth* (Garden City, N. Y.: Doubleday, 1955), pp. 137 ff., 191 ff. Cf. Paul Hanly Furfey, *The Scope and Method of Sociology: A Metasociological Treatise* (New York: Harper, 1953), pp. 38 ff.

4. Especially noteworthy among nonbehavioralists are the influential writings of Herbert Blumer on public opinion and collective behavior, and those of Leo Lowenthal on mass communication—the former falling broadly within an American tradition identified with G. H. Mead, C. H. Cooley and W. I. Thomas; the latter within the classic tradition of European social science. For discussion of these traditions, see below, Chap. VII.

5. The intimate relationship of this new field to sociology and social psychology is emphasized in Heinz Eulau, Samuel Eldersveld, and Morris Janowitz (eds.), *Political Behavior: A Reader in Theory and Research* (Glencoe, Ill.: Free Press, 1956), p. 3: "[The political behavior approach] seeks to place political theory and research in a frame of reference common to that of social psychology, sociology, and cultural anthropology." These authors seem, in fact, prepared to abandon the traditional discipline and designation of "political science" in favor of interdisciplinary merger: "Indeed, it could be argued that from a general social science point of view, political science (or, for that matter, sociology or social psychology) does not have a rationale for independent existence, and that all these disciplines might be merged into a single scientific approach to human affairs or social relations without detriment to the study of the specifically political aspects of behavior." *Ibid.*, p. 4. Cf. Evron M. Kirkpatrick, "The Impact of the Behavioral Approach on Traditional Political Science," in Austin Ranney (ed.), *Essays on the Behavioral Study of Politics* (Urbana: University of Illinois Press, 1962), p. 12.

6. See (in addition to Eulau, Eldersveld, and Janowitz, above) the statement of David Truman: "The latter-day rebels have mostly rallied around the banner of 'political behavior' to do battle with 'institutionalists.' Like most embattled revolutionists, many of them have unwisely and impetuously consigned to oblivion all of the works of their predecessors." Truman, "The Impact on Political Science of the Revolution in the Behavioral Sciences," in *Research Frontiers in Politics and Government* (Washington: Brookings Institution, 1955), pp. 215–216.

7. Bernard Berelson, "The Study of Public Opinion," in Leonard D. White (ed.), *The State of the Social Sciences* (Chicago: University of Chicago Press, 1956), pp. 304–305.

8. Robert Bierstedt, *The Social Order* (New York: McGraw-Hill, 1957), p. 20. Cf. Arnold Green, *Sociology* (New York: McGraw-Hill, 1956), pp. 8, 10: "At no time does the scientific investigator state his own opinion of what ought to be, or of what is right and what is wrong. His own biases and preferences do not enter the picture. . . . The scientist's job is to analyze, to explain, and to increase the store of knowledge. His job should not be confused with the wielding of political or economic power, power which he does not possess." See also E. B. Reuter, *Handbook of Sociology* (New York: Dryden, 1941), p. 12.

9. Nearly 30 years ago Mannheim remarked that "typical American [sociological] studies start from questions in nowise connected with those problems which arouse our passions in everyday political and social struggle. . . . as soon as political and social problems impose themselves we notice an immense reserve, a lack of social atmosphere." *Essays on Sociology and Social Psychology*, p. 191.

10. For a somewhat different perspective on the "pure vision of

the behavioral scientist," which "takes us far out of the world" of political reality, see Mulford Q. Sibley, "The Limitations of Behavioralism," in *The Limits of Behaviorialism in Political Science*, a symposium sponsored by The American Academy of Political and Social Science, edited by James C. Charlesworth (Philadelphia: The American Academy, October, 1962), pp. 87 ff.

11. Pauline V. Young, *Scientific Social Surveys and Research* (Englewood Cliffs, N. J.: Prentice-Hall, 3d ed., 1956), p. 123. It was presumably such ascetic inhibitions as these which led Louis Wirth to make his vigorous appeal for a reexamination of methodological assumptions: "The great unanswered questions of the social sciences are the great unanswered questions of mankind. How can we get peace, freedom, order, prosperity and progress under different conditions of existence? . . . I know these are general and cosmic questions, but until social scientists make a usable answer to the ways and means of achieving such ends, they will be playing a game which may be interesting enough to themselves, but one which they have no right to expect society to support." Quoted in Stuart Chase, *The Proper Study of Mankind* (New York: Harper, 1948), p. 56.

12. Cf. C. Wright Mills, *The Sociological Imagination* (New York: Oxford University Press, 1959), chap. 3, "Abstracted Empiricism," pp. 50 ff.

13. Examples are Talcott Parsons, *The Social System* (Glencoe, Ill.: Free Press, 1951); Marion J. Levy, *The Structure of Society* (Princeton: Princeton University Press, 1952); David Easton, *The Political System* (Chicago: University of Chicago Press, 1954). For criticism of the first two, see Barrington Moore, Jr., *Political Power and Social Theory* (Cambridge: Harvard University Press, 1958), chap. 3, "The New Scholasticism and the Study of Politics," pp. 89 ff. Cf. Ralf Dahrendorf, "Out of Utopia: Toward a Reorientation of Sociological Analysis," *The American Journal of Sociology*, Vol. LXIV, No. 2 (September, 1958), pp. 115–127. See also Mills, *The Sociological Imagination*, chap. 2, "Grand Theory," pp. 25–49.

14. Hans J. Morgenthau, "Power as a Political Concept," in Roland Young (ed.), *Approaches to the Study of Politics* (Evanston: Northwestern University Press, 1958), p. 70. Cf. the comment of John P. Roche: "I object to the spell of numerology which seems to have fallen over the study of politics, and my objections do not run against the techniques but against the basic assumptions of the new Establishment. . . . What I object to is not quantification, but bogus quantification, the assumption that one can create a measurable thing merely by assigning it a numerical symbol. No one in his senses doubts that we can tabulate Republican votes, but does it follow that we can count Oedipus complexes on the same easy basis? The first calculation involves counting heads; the second, measuring their content." "Political Science and Science Fiction," *The American Political Science Review*, Vol. LII, No. 4 (December, 1958), p. 1028.

15. Thus Morgenthau speaks of the "tendency, common to all methodological endeavors in the social sciences, to retreat ever more from contact with the empirical world into a realm of self-sufficient abstractions. . . . The new scholastic tends to think about how to think and how to conceptualize about concepts, regressing ever further from empirical reality until he finds the logical consummation of his endeavors in mathematical symbols and other formal relations." "Power as a Political Concept," p. 70.

16. H. J. Eysenck, *The Psychology of Politics* (New York: Praeger, 1954), dedication page.

17. This is, of course, an old distinction, perhaps first articulated in its modern form by Max Weber. See, for example, his *The Methodology of the Social Sciences,* translated and edited by E. A. Shils and H. A. Finch (Glencoe, Ill.: Free Press, 1949), esp. chap. I and II. For a detailed and decisive critique of Weber's value-neutralism, see Leo Strauss, *Natural Right and History* (Chicago: University of Chicago Press, 1953), chap. 2, "Natural Right and the Distinction Between Facts and Value," pp. 13–20. Another source of the Weber position is to be seen in Max Weber, "Politics as a Vocation," in Hans Gerth and C. Wright Mills (eds.), *From Max Weber: Essays in Sociology* (London: Routledge and Kegan Paul, 1948), pp. 77–128. Weber's other side, that of *Verstehen,* with its vastly different implications, is noted in Chap. 7, below.

18. Eysenck, *The Psychology of Politics,* p. 2. A sociologist writes in a similar spirit: "Once a person accepts a value, he becomes somewhat biased or prejudiced thereby. This does not mean that all people who accept the value of democracy are necessarily *equally* biased or prejudiced, but merely that they are all likely to be somewhat blinded to the possible values of a non-democratic social system. Similarly one may prefer rural life to city life, the Republican Party to the Democratic Party. Thus, one can not escape having some values, and once he has them they operate as prejudices or biases in his thinking." John F. Cuber, *Sociology: A Synopsis of Principles* (New York: Appleton-Century, 1947), p. 21.

19. George A. Lundberg, *Can Science Save Us?* (New York: Longmans, Green, 1947), p. 27. Lundberg is generally regarded as leader of the "natural science" school in sociology, whose members include Stuart Dodd and Read Bain. See, for example, G. A. Lundberg, "Contemporary Positivism in Sociology," *American Sociological Review,* Vol. IV (1939), pp. 42–55; Lundberg, "Operational Definitions in the Social Sciences," *American Journal of Sociology,* Vol. XLVII (1942), pp. 727–743; Read Bain, "Sociology as a Natural Science," *American Journal of Sociology,* Vol. LIII (1947), pp. 10 ff. For critical analyses of Lundberg's position, see Furfey, *The Scope and Method of Sociology,* pp. 38 ff.; Timasheff, *Sociological Theory,* pp. 191 ff.; George Simpson, "The Assault on Social Science," *American Sociological Review,* Vol. XIV (1949), pp. 303–310.

20. Thus Lundberg goes on to say: "The confusion about values

seems to have arisen because both scientists and the public have frequently assumed that, when scientists engage in ordinary pressure-group activity [sic], that activity somehow becomes science or scientific activity. This is a most mischievous fallacy. . . . It is unpardonable for scientists themselves to be confused about what they know and say in their capacity as scientists and what they favor in religion, morals, or public policy. To pose as disinterested scientists announcing scientific conclusions when in fact they are merely expressing personal preferences is simple fraud." *Can Science Save Us?* pp. 27–28.

21. *Ibid.*, p. 39.

22. *Ibid.* On this *hubris* of the pollsters, Peter H. Odegard has written: "To say that the polls made possible direct democracy on a continental basis was to give aid and comfort to those who attacked them as subversive of representative democracy. To say that the polls might even be better than elections for recording the general will on public issues was construed as lack of faith in the American Way. . . . Isn't it time for the pollsters to cease confusing verbal reactions of a selected sample of people to hypothetical questions with the voice of God and themselves as high priests?" *American Political Science Review*, Vol. XLIV, No. 2 (June, 1950), pp. 461–463. The classic demolition of polling pretensions is Lindsay Rogers, *The Pollsters: Public Opinion, Politics, and Democratic Leadership* (New York: Knopf, 1949), esp. Books I, II and V.

23. *Can Science Save Us?* pp. 43–44.

24. *Ibid.*, p. 44. On "aggression," Lundberg remarks: "To take only one example, 'statesmen,' preachers, and journalists have recently frothed at the mouth about the aggressions of Germany. It would be interesting to see an objective definition of aggression which will not apply with equal validity to Russia's operations. . . . (I pick on Russia for this illustration merely because mention of the world depredations of the United States and Britain would grate unduly upon the lofty sensibilities of these world liberators.) To have to face such facts is undoubtedly one of the greatest costs of a transition to a scientific view of human relations." *Ibid.* Apparently, from a scientific view, all acts of aggression are of equal weight, and none is worth frothing over.

25. *Ibid.* "Social sciences worthy of the name will have to examine realistically all the pious shibboleths which are not only frequently the last refuge of scoundrels and bigots, but also serve as shelters behind which we today seek to hide the facts we are reluctant to face. The question is, how much pain in the way of disillusionment about fairy tales, disturbed habits of thought, and disrupted traditional ways of behavior will the patient be willing to put up with in order to be cured of his disease?" *Ibid.*, p. 45.

26. *Ibid.*

27. *Ibid.* (Emphasis added.)

28. *Ibid.*, p. 46.

29. See his statement on pp. 51–54, as well as the following observation: "When people are in trouble, they will look for a saviour. . . . They are likely to surround themselves with seers, poets, playwrights, and others alleged to possess these powers of 'seeing.' The idea is a sound one. The only reform needed is a substitution of scientists for these soothsayers and soothseers." *Ibid.*, p. 54.

30. *Ibid.*, p. 55.

31. "A similar attitude toward the conclusions of social scientists is suspected of being authoritarian, as indeed it probably is. A lot of nonsense has been spoken and written about authority in recent years. We need to recognize that it is not authority as such that we need fear but incompetent and unwisely constituted authority. When we undertake to insist on the same criteria of authority in the social as in the physical sciences, no one will worry about the delegation of that authority, any more than he worries about the physician's authority." *Ibid.*, p. 51.

32. *Ibid.*, pp. 47–48. (Emphasis added.)

33. *Ibid.*, p. 97.

34. See especially Reinhard Bendix, *Social Science and the Distrust of Reason* (Berkeley: University of California Press, 1951), pp. 37 ff. Cf. Mills, *The Sociological Imagination*, pp. 95 ff., 100 ff.; Dwight Waldo, *The Study of Public Administration* (Garden City, N. Y.: Doubleday, 1955), p. 64.

35. See especially Morton Grodzins, "Public Administration and the Science of Human Relations," *Public Administration Review*, Vol. VII (1951), pp. 88–102. "The short point is that the science of human relations constitutes an effective tool for the manipulation of men. A very large portion of scientific knowledge about human relations is the result of research geared to manipulative purposes. The Western Electric Company paid Mayo and his collaborators to increase the productivity of workers. Stouffer and his colleagues and Grinker and Spiegel had direct responsibilities for adding to the fighting qualities of men in the armed services. Leighton's work was done as an adjunct to the wartime Japanese evacuation and the larger purpose was to provide guidance to the military for the administration of conquered peoples. So the list goes." *Ibid.*, p. 100. Cf. Bendix, *Social Science and the Distrust of Reason, passim.*; William H. Whyte, Jr., *The Organization Man* (New York: Doubleday Anchor Books, 1957), pp. 25 ff.

36. Gunnar Myrdal, *An American Dilemma: The Negro Problem and Modern Democracy* (New York: Harper, 1944).

37. George A. Lundberg, "The Proximate Future of American Sociology: The Growth of Scientific Method," *American Journal of Sociology*, Vol. L (1945), pp. 502–513.

38. "A Methodological Note on Facts and Valuations in Social Science," Appendix Two, *An American Dilemma*, pp. 1035 ff. Cf. Furfey, *The Scope and Method of Sociology*, p. 45n.

39. Some words of Lundberg's, quoted earlier, find an echo here. After observing in effect that social scientists should strive to become so passively accommodating as to be deemed indispensable by any regime, he points out that a few among them have already approximated to this ideal: namely, "qualified social statisticians [who] have not been and will not be disturbed greatly in their function by any political party as long as they confine themselves to their specialty." The inference would seem to be that what most scientists have traditionally regarded as a routine technical facility is properly to be viewed as the pinnacle of social-scientific achievement.

40. On the contrary, it almost seems from the behaviorist definition of science that everything human is alien to it; that "science" is an activity not of living beings but of pure spirit (allied with pure number), inhabiting a realm of essences beyond good and evil, and disengaged not only from the sinful flesh but also from the earthbound mind and purposes of man. The evangelism which lies behind the rhetorical question "Can Science Save Us?" is evidently not a call to arms against the evils and injustices of this world, but an appeal to rise above them through the attainment of proficiency in something resembling a new mystique of nonattachment. Cf., on "heuristic passion" and commitment to value in natural science, Michael Polanyi, *Personal Knowledge* (Chicago: University of Chicago Press, 1959), pp. 142 ff., *passim*. See also, for vivid illustrations of "creative confusion" among the great discoverers, Arthur Koestler, *The Sleepwalkers* (New York: Macmillan, 1959); W. I. B. Beveridge, *The Art of Scientific Investigation* (New York: Norton, n.d.); J. Bronowski, *Science and Human Values* (New York: Messner, 1956); David Lindsay Watson, *The Study of Human Nature* (Yellow Springs, Ohio: Antioch Press, 1953).

41. "[Science] consists of '(a) asking clear, answerable questions in order to direct one's (b) observations, which are made in a calm and unprejudiced manner, and which are then (c) reported as accurately as possible and in such a way as to answer the questions that were asked to begin with, after which (d) any pertinent beliefs or assumptions that were held before the observations were made are revised in light of the observations made and answers obtained.'" *Can Science Save Us?* p. 67. Lundberg notes that he has taken this definition from Wendell Johnson, a general semanticist of note.

42. *Modern Science and Modern Man* (New York: Doubleday Anchor Books, 1952), p. 119.

43. *Ibid.* The pervasive role of values and value-judgments in scientific inquiry is emphasized by Conant throughout this book.

44. Cf. Bernard Barber, *Science and the Social Order* (Glencoe,

Ill.: Free Press, 1952); J. D. Bernal, *The Social Functions of Science* (New York: Macmillan, 1939). The classical statement of the sociology of knowledge is Karl Mannheim, *Ideology and Utopia* (New York: Harcourt, Brace, 1936). Mannheim's work is discussed further in Chap. VII, below.

45. Max Lerner, *America as a Civilization* (New York: Simon and Schuster, 1957), p. 236.

46. *Ibid.* Cf. Erich Kahler, *Man the Measure* (New York: Pantheon, 1943), pp. 567 ff.

47. Cf. Erich Fromm, *Man for Himself* (New York: Rinehart, 1947), pp. 67 ff., *passim;* David Riesman, *The Lonely Crowd* (New Haven: Yale University Press, 1959), chap. 6, 7; C. Wright Mills, *White Collar* (New York: Oxford University Press, 1951), chap. 5.

48. Herbert Kaufman, "Emerging Conflicts in the Doctrines of Public Administration," *American Political Science Review*, Vol. L, No. 4 (December, 1956), p. 1060.

49. Dwight Waldo, *The Study of Public Administration* (Garden City, N. Y.: Doubleday, 1955), p. 41. Cf. Waldo, *The Administrative State* (New York, 1948), chap. 3, "Scientific Management and Public Administration."

50. "Emerging Conflicts in the Doctrines of Public Administration," p. 1060.

51. "Neutral competence is still a living value among students of government, career civil servants, and, perhaps more significantly, among much of the general populace." *Ibid.,* p. 1062.

52. Waldo, *The Study of Public Administration,* pp. 42 f. Cf. Waldo, *Ideas and Issues in Public Administration: A Book of Readings* (New York: McGraw-Hill, 1953), pp. 407–408. See, for a book-length illustration, Paul M. Appleby, *Policy and Administration* (Birmingham: University of Alabama Press, 1945).

53. "Emerging Conflicts in the Doctrines of Public Administration," p. 1070.

54. *Ibid.,* p. 1072. Cf. Waldo, *The Administrative State*, chap. 6, "Who Should Rule."

55. Even the roles are not distinct in the case of the behavioral scientist seeking, like Bernard Berelson, to become a member of a "team of technicians." For acid commentary on this trend, see Mills, *The Sociological Imagination*, chap. 5, "The Bureaucratic Ethos."

56. The pertinent body of work by Lasswell is cited below. Another prominent example is Herbert A. Simon, *Administrative Behavior* (New York: Macmillan, 1947).

57. More mystifying still is Lasswell's coinage of the "value sci-

ences" as a synonym for the social sciences. Harold D. Lasswell and Abraham Kaplan, *Power and Society: A Framework for Political Inquiry* (New Haven: Yale University Press, 1950), p. xxiv. Cf. Harold D. Lasswell and Daniel Lerner (ed.), *The Policy Sciences: Recent Developments in Scope and Method* (Stanford: Stanford University Press, 1951). See also Simon, *Administrative Behavior*, chap. IX, *passim*.

58. Little purpose would be served by citing the standard and generally familiar works in this field (by such writers as Carnap, Neurath, Reichenbach, Philip Frank, Ayer, Bergman, Richard von Mises, *et al.*). More immediately relevant to political science is the massive exegesis of modern political theory, from the standpoint of "Scientific Value Relativism," by Arnold Brecht: *Political Theory: The Foundations of Twentieth Century Political Thought* (Princeton: Princeton University Press, 1959). Other sympathetic applications or interpretations within political science include Simon, *Administrative Behavior* (2nd ed., 1957), Introduction; T. D. Weldon, *The Vocabulary of Politics* (London: Penguin Books, 1953). Recent critical accounts include Leo Strauss, *Natural Right and History*, Introduction, chap. II; Pitirim Sorokin, *Fads and Foibles in Modern Sociology and Related Sciences* (Chicago: Regnery, 1956), chap. 2, 12, *passim;* Michael Polanyi, *The Logic of Liberty* (Chicago: University of Chicago Press, 1951), pp. 43 ff., 62 ff.; Lee Cameron McDonald, "Voegelin and the Positivists: A New Science of Politics?" *Midwest Journal of Political Science*, Vol. I, No. 3–4 (November, 1957), pp. 233–251.

59. Cf. Harold D. Lasswell, *The Analysis of Political Behavior: An Empirical Approach* (London: Routledge and Kegan Paul, 1948) pp. 2–3, 12. See also Lasswell and Kaplan, *Power and Society*, p. x: "Theorizing, even about politics, is not to be confused with metaphysical speculation in terms of abstractions hopelessly removed from empirical observation and control."

60. Thus Lasswell and Kaplan soberly report the results of an unusual bit of content analysis: "A rough classification of a sample of 300 sentences from each of the following yielded these proportions of political philosophy (demand statements and valuations) to political science (statements of fact and empirical hypotheses): Aristotle's *Politics*, 25 to 75; Rousseau's *Social Contract*, 45 to 55; Laski's *Grammar*, 20 to 80. Machiavelli's *Prince*, by contrast, consisted entirely (in the sample) of statements of political science in the present sense." *Power and Society*, p. 118n.

61. See, for example, Lundberg, *Can Science Save Us?* pp. 61 ff., *passim*. Cf. the rather impatient "noncognitivism" of Felix Oppenheim, "The Natural Law Thesis: Affirmation or Denial," *American Political Science Review*, Vol. LI (1957), pp. 41–53.

62. Simon, *Administrative Behavior*, chap. 9, *passim;* Lasswell, *The Analysis of Political Behavior*, pp. 193–194: "It would be possible to make an extensive analysis from the tripartite point of view of the meaning

of social doctrines, myths and legends. Some popular sayings appeal to the ego: 'Doubt is the beginning of wisdom.' Some appeal to the conscience: 'Honour thy father and thy mother.' Some appeal to the id: 'You're only young once.' Stories of heroic and villainous acts, prophecies of heroic or villainous events, *theories* of social permanence or social flux—all combine to arouse different components of the personality, stimulating the ego, reinforcing the super-ego, and unleashing the id."

63. *The Study of Public Administration*, p. 44. The reference is primarily to the "logical-positivist" approach exemplified in public administration by Herbert Simon.

64. See Lasswell and Kaplan, *Power and Society*, pp. 118 ff.

65. "To rescue science from this hopeless gambit, three modern developments have converged on a common solution. The three are behaviorism in psychology, operationism in physics, and logical positivism in philosophy. . . . all three of these movements have sought to clarify our scientific discourse by ridding its concepts of metaphysical overtones and untestable meanings." S. S. Stevens, "Measurement and Man," *Science*, Vol. CXXVII (1958), p. 386. The "physicalism" of Otto Neurath and Rudolf Carnap in the 1930's bore obvious resemblances to psychological behaviorism in its rejection of introspective sources of knowledge and its linguistic austerity. Cf. Brecht, *Political Theory*, pp. 174 ff.

66. See Harold D. Lasswell, "Politics: Who Gets What, When, How," in *The Political Writings of Harold D. Lasswell* (Glencoe, Ill.: Free Press, 1951), pp. 386 ff.; Lasswell and Kaplan, *Power and Society*, pp. 118 ff.

67. On Freud, see below, Chap. VII, "Freud: Romantic Mechanist."

68. On the influence of Lasswell, see David Easton, "Harold Lasswell: Policy Scientist for a Democratic Society," *Journal of Politics*, Vol. XII (1950), pp. 450–477; Heinz Eulau, "H. D. Lasswell's Developmental Analysis," *Western Political Quarterly*, Vol. XI, No. 2 (1958), pp. 229–242.

69. Lasswell, *World Politics and Personal Insecurity* (Chicago: University of Chicago Press, 1934), pp. 22–23, 135–136.

70. See the discussion of this point in Bernard Crick, *The American Science of Politics: Its Origins and Conditions* (Berkeley: University of California Press, 1959), pp. 182 ff.

71. *World Politics and Personal Insecurity*, pp. 22–24.

72. "Politics," p. 295.

73. *World Politics and Personal Insecurity*, p. 3.

74. "Politics," p. 295. Cf. *Power and Society*, p. xi.

75. *Analysis of Political Behavior*, p. 181. See also *ibid.*, p. 8.

76. "Politics," pp. 306–307. Cf. *Power and Society*, p. xi.

77. *Analysis of Political Behavior*, p. 134.

78. *World Politics and Personal Insecurity*, p. 237. Thus the Grand Inquisitor of Dostoevski's *Brothers Karamazov*, discussing an eminently successful world myth of nonrational consensus: ". . . we too have a right to preach a mystery, and to teach them that it's not the free judgment of their hearts, not love that matters, but a mystery which they must follow blindly even against their conscience. So we have done. We have corrected Thy work and have founded it upon miracle, mystery, and authority."

79. "Politics," p. 304. Cf. "Psychopathology and Politics," in *Political Writings*, pp. 52 f., *passim*.

80. "Politics," p. 298.

81. See, for example, the critique of Mannheim by Karl R. Popper, *The Open Society and Its Enemies* (London: Routledge and Kegan Paul, 1945), Vol. II, chap. 23.

82. *Ibid.*, p. 390.

83. *World Politics and Personal Insecurity*, p. 20.

84. This interest of Lasswell's, like his interest in elites, has been subject to shifts in perspective but has remained close to the center of his attention. Cf. Lasswell, "The Impact of Psychoanalytic Thinking on the Social Sciences," in L. D. White (ed.), *The State of the Social Sciences* (Chicago: University of Chicago Press, 1956); "The Normative Impact of the Behavioral Sciences," *Ethics*, Vol. LXVII, No. 3 (April, 1957), pp. 26 ff.; *Power and Personality* (New York: Norton, 1948); "Democratic Character," in *Political Writings*, pp. 480 ff.

85. "Psychopathology and Politics" (1930), p. 197.

86. *Ibid.*, pp. 196–197.

87. *Ibid.*, p. 194.

88. *Ibid.*, pp. 202, 194.

89. If Madison would have taken issue with this solution, it is likely that Freud himself would have been little more enthusiastic; for he also regarded "factional" instincts and passions as inherent in the nature of man, and warned that society represses them at its peril. See his *Civilization and Its Discontents* (London: Hogarth Press, 1930), esp. pp. 65, 86 ff. On the contemporary implications of Madison's pluralism (as stated in the *Federalist*, No. 10), see Floyd W. Matson, "Party and Faction: The Principles of Politics vs. the Politics of Principle," *Antioch Review*, Vol. XVIII, No. 3 (Fall, 1958), pp. 331–342.

90. "Psychopathology and Politics," p. 197.

91. "If the politics of prevention spreads in society, a different type of education will become necessary for those who administer society or think about it. . . . The social administrator and social scientist must be brought into direct contact with his material in its most varied manifesta-

tions." (*Ibid.*, p. 201.) "The achievement of the ideal of preventive politics depends less upon changes in social organization than upon improving the methods and the education of social administrators and social scientists." *Ibid.*, p. 203.

92. *Ibid.*, p. 196.

93. Cf. Lasswell, Nathan Leites and associates, *The Language of Politics: Studies in Quantitative Semantics* (New York: 1949); Lasswell, Daniel Lerner and Ithiel de Sola Pool, *The Comparative Study of Symbols* (Stanford: Stanford University Press, 1952); Lasswell, "The World Attention Survey," in *Analysis of Political Behavior*, pp. 296–303.

94. For discussion by a number of specialists of the applications of content analysis to such fields as political science, sociology, psychology, psychotherapy, history, linguistics, and folklore, see Ithiel de Sola Pool (ed.), *Trends in Content Analysis* (Urbana: University of Illinois Press, 1959). For a critique of the depth and relevance of such studies, cf. Crick, *The American Science of Politics*, pp. 187–188.

95. "At Chicago and Yale, Professor H. D. Lasswell has used an admirable technique, adapted from the economists. It is a species of political logical positivism and involves a study of semantics." George Catlin, "The Utility of Political Science," in Morroe Berger, T. Abel, and C. H. Page, *Freedom and Control in Modern Society* (New York: Van Nostrand, 1954), p. 266. Catlin subsequently adds: ". . . what we require is not 'no ideology,' but a better ideology and one very carefully considered—at least as much as Marxism, the product of a century of brain work. We need a political logical positivism, a new political scholasticism, for a new world." *Ibid.*, p. 282.

96. On this and other aspects of logical positivism, and its variant of linguistic philosophy, see the formidable critique by Ernest Gellner, *Words and Things* (Boston: Beacon Press, 1959), which also carries an introduction by a reconstructed godfather of the school: Bertrand Russell. Cf. McDonald, "Voegelin and the Positivists," cited previously.

97. At the hands of systematic exponents, this becomes the tendency to classify and measure in quantitative terms. Thus Lasswell and Kaplan: "Rather than isolating democracy, for instance, as an absolutely distinct political form, we specify a range of democratic-despotic characteristics. . . . Once concepts are formulated in terms of order, of variation in degree, scales and methods of measurement can be developed to suit the needs of the particular problem." *Power and Society*, pp. xvi–xvii. Cf. the effort at "quantification of ethics" by Anatol Rapoport, *Operational Philosophy* (New York: Harper, 1954), pp. 190–192.

98. The classic attempt to apply such methods to literary criticism is I. A. Richards, *The Principles of Literary Criticism* (London: 1924). Cf. Polanyi, *The Logic of Liberty*, p. 9.

99. The ambiguous term is that of Catlin; see Note 95, above.

100. *Power and Society*, p. xi.

101. *Ibid.*, p. xiii.

102. *Ibid.*

103. "This standpoint may be designated, if a term is wanted, as *hominocentric politics*. As science, it finds its subject matter in interpersonal relations, not abstract institutions or organizations; . . . As policy, it prizes not the glory of a depersonalized state or the efficiency of a social mechanism, but human dignity and the realization of human capacities." *Ibid.*, p. xxiv.

104. *Ibid.*

105. *Ibid.*, p. xiii.

106. *Ibid.* The term "fruitful" may be passed by as merely redundant in the context; but the use of the term "warranted," by social scientists who are professedly unconcerned with the "justification" or "derivation" of "moral values," emphasizes the dilemma broadly under discussion here. In such a system as theirs there is no warrant for "warrantability," but there is nevertheless both an assumption of it and an appeal to it. The impasse is not peculiar to Lasswell and Kaplan; it is presented most prominently by the late T. D. Weldon's *Vocabulary of Politics*. In an incisive critique of Weldon, Joseph Margolis concludes that this "is the question that we have tracked through Weldon's entire text and found unexamined. And it is, in fact, one of the most crucial questions that moral, legal, and political philosophy have yet to address themselves to in a convincing and fruitful way." "Difficulties in T. D. Weldon's Political Philosophy," *American Political Science Review*, Vol. LII, No. 4 (December, 1958), p. 1117.

107. *Power and Society*, pp. xiii–xiv.

108. See Lasswell, *Democracy Through Public Opinion* (New York: 1945), p. 7; *Power and Personality*, p. 107; *Analysis of Political Behavior*, p. 2; "Democratic Character," pp. 471, 473; *Policy Sciences*, p. 15; *The World Revolution of Our Times: A Framework for Basic Policy Research* (Stanford: Stanford University Press, 1951), p. 8; *Power and Society*, p. xiii.

109. *Analysis of Political Behavior*, p. 7. Elsewhere Lasswell has noted that the special science of democracy is only one possibility within the rubric of policy sciences: "There can also be a policy science of tyranny. A science of total policy would select either set of postulates and analyze both." ("Democratic Character," p. 471n.) The selection would, however, be utterly capricious—unless such a "total" science should recognize an affinity with a total(itarian) system of politics.

110. "Democratic Character," p. 474. Cf. *The World Revolution of Our Times*, Appendix A.

111. "Democratic Character," p. 474.

112. William Ernest Hocking, *Experiment in Education* (Chicago: Regnery, 1954), pp. 228, 238 ff. Lasswell has indeed spoken of the "working specification of goal-values," but only in the empirical and pragmatic sense of seeking objective correlates of such values—which may then be shaped, shared and allocated. His evident reification of "dignity" and "freedom" comes very close to illustrating Whitehead's fallacy of misplaced concreteness. In a brilliant review of *The Policy Sciences*, Carl J. Friedrich concludes: ". . . if Lasswell builds his entire approach upon the proposition that 'the special emphasis is upon the policy sciences of democracy, in which the ultimate goal is the realization of human dignity in theory and fact,' the student of public policy is entitled to explore the answer to such questions as: what constitutes an ultimate social goal, what is the meaning of democracy, of dignity, and last but not least of man?" "Policy—A Science?" in C. J. Friedrich and J. K. Galbraith, *Public Policy* (Cambridge: Harvard University Graduate School of Public Administration, 1953), Vol. IV, p. 281.

113. Lasswell, "Conflict," *Encyclopedia of the Social Sciences* (New York: 1930), Vol. V, pp. 194–196.

114. *Analysis of Political Behavior*, p. 17.

115. *World Revolution of Our Times*, p. 8.

116. For an elaboration of this argument, see Floyd W. Matson, "Social Welfare and Personal Liberty: The Problem of Casework," *Social Research*, Vol. XXII, No. 3 (Autumn, 1955), pp. 253–274. Cf. Jacobus tenBroek and Floyd W. Matson, *Hope Deferred: Public Welfare and the Blind* (Berkeley: University of California Press, 1959), chaps. 2, 8.

117. This is the purport of "The Normative Impact of the Behavioral Sciences," in which that impact is found to be uniformly affirmative and democratic. But this finding displays, in addition to selective perception in the examples and illustrations chosen, an unmistakable degree of selective inattention as well. Thus Lasswell points to "a remarkable contrast" in the application of the physical and behavioral sciences: "It is the physical sciences and technologies that have been heavily exploited by the antidemocratic elements of civilization. *Remarkably little use has been made, by these forces, of behavioral knowledge.*" (P. 36.) This is seen to follow from the fact that physical science is basically "coercive," while behavioral science is "persuasive"—and persuasion is surely more democratic than coercion. "There appears to be a sense, therefore, in which democratic processes are more intimately tied up with the behavioral sciences than with other kinds of scientific knowledge." (P. 37.) The only comment that need be made concerning this argument is that its curious blindness toward the authoritarian uses of "persuasion" (known to every reader of *Brave New World* and *Nineteen Eighty-Four*) is not altogether inappropriate for a political scientist to whom manipulation is the practical counter-

part of contemplation. For a detailed catalogue of behavioral-scientific contributions to "tyranny over the mind," see Aldous Huxley, *Brave New World Revisited* (New York: Harper, 1958).

118. See Crick, *The American Science of Politics*, esp. Parts III and IV. Cf. the editorial note by Charlotte Luetkens in Lasswell, *Analysis of Political Behavior*, p. viii: "The book, therefore, remains an American book: American in background, in the enviable amplitude of resources and scholarly teamwork [sic], American in the acute awareness of slipping standards and ever-encroaching techniques."

119. Graven in stone over the entrance of the social science research building at Lasswell's own University of Chicago are the words of the physicist Lord Kelvin: "Only that is knowledge which can be weighed and measured." (It was also Kelvin, as noted in Chapter IV, below, who announced that if there was one thing of which we could be certain in physics it was the reality of the luminiferous ether.)

It has been maintained, by David Easton and others, that the evolution of Lasswell's democratic commitment has meant the decline or disappearance of his earlier elitism. Lasswell has, it is true, made such statements as that "the elite of democracy ('the ruling class') is society-wide" (*Power and Personality*, p. 108); but it is no less true that his recent writings have been, as Arnold Brecht points out, "replete with 'elites,' with no mention of the people." A striking example is Lasswell's APSA Presidential Address of 1956, "The Political Science of Science," which is notable not only for its familiar conjuring of "elite, mid-elite, and mass" but still more so for its earnest admonition to political scientists to prepare themselves for the imminent reception of strange new "masses" and "elites": "When do we accept the humanoids—the species intermediate between lower species and man, . . . as at least partial participants in the body politic? And at what point do we accept the incorporation of relatively self-perpetuating and mutually influencing 'super-machines' or 'ex-robots' as beings entitled to the policies expressed in the Universal Declaration [of Human Rights]? . . . It is plain that if we bring certain kinds of living forms into the world we may be introducing a biological elite capable of treating us in the manner in which imperial powers have so often treated the weak." *American Political Science Review*, Vol. L, No. 4 (December, 1956), pp. 976–977. It cannot, after this, be charged against Lasswell that he has neglected his own prescription for political analysis: i.e., "the imaginative charting of new possibilities."

120. See William Y. Elliott, *The Pragmatic Revolt in Politics* (New York: Macmillan, 1928); Morton G. White, *Social Thought in America: The Revolt Against Formalism* (New York: Viking, 1949); Louis Hartz, *The Liberal Tradition in America* (New York: Harcourt, Brace, 1955).

121. See Arthur F. Bentley, *Behavior, Knowledge, Fact* (Bloomington, Ind.: Principia Press, 1936), *passim*. (And see below, Note 124.)

122. This rejection was a remarkable about-face from his earliest methodological standpoint. His doctoral dissertation, taken at Johns Hopkins in 1896, was entitled "The Units of Investigation in the Social Sciences." As Ratner has described the essay, it "stressed the human mind as a central point for all study of social phenomena, and reflected mentalistic or psychical formulations then current which Bentley later came to discard completely." Sidney Ratner, Introduction to Arthur F. Bentley, *Inquiry into Inquiries: Essays in Social Theory* (Boston: Beacon Press, 1954), p. x.

123. "The individual stated for himself, and invested with an extra-social unity of his own, is a fiction." Arthur F. Bentley, *The Process of Government: A Study of Social Pressures* (Bloomington, Ind.: Principia Press, 1949 edition), p. 215. The main development of Bentley's group interpretation is to be found in *ibid.*, chap. 7, "Group Activities."

124. Cf. Bentley, *Inquiry into Inquiries*, chap. 3, "A Sociological Critique of Behaviorism." The essay is less a critique than an extension of Watson's precepts, embracing "a wholehearted agreement with Watson" in such particulars as "the rejection of all terms dealing with consciousness or mentality in any of its technical forms, and the refusal to use categories arising from terms of this type for any scientific purpose." *Ibid.*, p. 32.

125. The main work is Gustav Ratzenhofer, *Die sociologische Erkenntnis* (Leipzig, 1898). Cf. Albion W. Small, *General Sociology* (Chicago: University of Chicago Press, 1905), esp. chap. 8; Harry Elmer Barnes, *Introduction to the History of Sociology* (Chicago: University of Chicago Press, 1947), chap. XIX.

126. See the provocative treatment of this theme in C. Wright Mills, *The Power Elite* (New York: Oxford University Press, 1959 Galaxy Edition), chap. 11, "The Theory of Balance."

127. See the prefatory pages appearing only in the 1935 edition of Bentley's *Process of Government*. Cf. Ratner's introduction to Bentley's *Inquiry into Inquiries*.

128. See especially pp. 26 ff., 468 ff.

129. For Bentley the "idea and feeling elements" were of no use to political science primarily because they were "unmeasureable as they appear in studies of government. This is a fatal defect in them." *Process of Government*, p. 201. Much of Bentley's early chapters were devoted to arguing, not very coherently, the case against "ideas" and "feelings," always rendered in quotation marks.

130. "As for political questions under any society in which we are called upon to study them, we shall never find a group interest of the society as a whole. . . . The society itself is nothing other than the complex of the groups that compose it." *Ibid.*, p. 222. Bentley's apparent vacillation on this score is noted by R. E. Dowling, "Pressure Group Theory: Its

Methodological Range," *American Political Science Review*, Vol. LIV, No. 4 (December, 1960), pp. 944–954.

131. See his discussion of "activity" as the social equivalent or expression of matter in motion, *Process of Government*, pp. 184 ff. Cf. Dowling, "Pressure Group Theory," pp. 944–946.

132. "Now, . . . the activities are all knit together in a system, and indeed only get their appearance of individuality by being abstracted from the system; they brace each other up, hold each other together, move forward by their interactions, and in general are in a state of continuous pressure upon one another." *Process of Government*, p. 218.

133. Raymond A. Bauer, *The New Man in Soviet Psychology* (Cambridge: Harvard University Press, 1952), p. 26.

134. "There is no political process that is not a balancing of quantity against quantity. There is not a law that is passed that is not the expression of force and force in tension." *Process of Government*, p. 202. Bentley goes on to argue, at some length, that "no group can be stated, or defined, or valued . . . except in terms of other groups. No group has meaning except in its relations to other groups." *Ibid.*, p. 217.

135. "Here we have been thinking for nearly three hundred years that if there is one causal law which is certain beyond challenge, it is the law of gravitation. The whole tradition of causality derives from its triumph. . . . And yet, and yet, the laws of gravitation have gone. There is no gravitation; there is no force at all, the whole model was wrong. All that theory was no more than a happy approximation to what really happens. When Newton brought in force as a cause, he was giving to matter the human property of effort, as much as Aristotle once gave it human will. The true causes are now embedded in the nature of space and the way in which matter distorts space; and they have no resemblance to the causes in which we believed for nearly three hundred years." J. Bronowski, *The Common Sense of Science* (Cambridge: Harvard University Press, 1955), pp. 64–65.

136. The famous comment of Eddington is pertinent here: "We have found that where science has progressed the farthest, the mind has but regained from nature that which the mind has put into nature. We have found a strange footprint on the shores of the unknown. We have devised profound theories, one after another, to account for its origin. At last, we have succeeded in reconstructing the creature that made the footprint. And Lo! It is our own." Quoted in Werner Heisenberg, *The Physicist's Conception of Nature* (New York: Harcourt, Brace, 1958), p. 153.

137. The phrase is that of Professor Arthur W. MacMahon, as quoted in Myron Q. Hale, "The Cosmology of Arthur F. Bentley," *American Political Science Review*, Vol. LIV, No. 4 (December, 1960), p. 958. This article is a cogent argument for the thesis that Bentley's cosmology

bears politically conservative implications, regardless of the "progressive" context in which Bentley carried on his work.

138. For a brilliant statement of the historical background, as well as a critique of shortcomings, see Peter H. Odegard, "A Group Basis of Politics: A New Name for an Ancient Myth," *Western Political Quarterly,* Vol. XI, No. 3 (September, 1958), pp. 689–702.

139. "There is a radical difference between being a contender for power, a rival among rivals, and being the guardian of the order which intends to regulate all the rivalries. In the one, the technique of the balance of power is used as an instrument of aggression and defense. In the other, it is used as the structural principle of public order in the good society." Walter Lippmann, *The Public Philosophy* (New York: Mentor Books, 1955), p. 122.

140. For an example of this inside-dopestership, on the part of a professed Bentleyan, see Bertram M. Gross, *The Legislative Struggle: A Study in Social Combat* (New York: McGraw-Hill, 1953).

141. The phrase, "rules of the game," was used casually by Bentley but has received its most explicit and systematic treatment in the updating of Bentley's theory by David B. Truman, *The Governmental Process* (New York: Knopf, 1955), as indicated by the numerous references in the index. An early journalistic expression is Frank R. Kent, *The Great Game of Politics* (New York: 1930). And, of course, there is the insistent "realism" of Lasswell's title, *Politics: Who Gets What, When, How.*

142. On the other hand, this sophistication (in the sense of logical and mathematical abstraction and articulation) is most marked on the part of those who have taken the game analogy most seriously: i.e., the exponents of "game theory." The present discussion does not attempt an analysis of this recent fashion in behavioral research, which presents problems of a separate and special kind. Some representative titles are: Anatol Rapoport, *Fights, Games, and Debates* (Ann Arbor: University of Michigan Press, 1960); M. Shubik, *Readings in Game Theory and Political Behavior* (Garden City, N. Y.: Doubleday, 1954); D. Luce and H. Raiffa, *Games and Decisions* (New York: Wiley, 1957); R. Braithwaite, *Theory of Games as a Tool for the Moral Philosopher* (New York: Cambridge University Press, 1955).

143. S. E. Finer, "In Defence of Pressure Groups," *The Listener* (June 7, 1956), p. 752.

144. Odegard, "A Group Basis of Politics," p. 699.

145. Michael Polanyi, *Personal Knowledge: Towards a Post-Critical Philosophy* (Chicago: University of Chicago Press, 1958), p. 380.

Part Two

Chapter IV *An Uncertain Trumpet*

1. George Santayana, *The Life of Reason* (New York: Scribner's, 1954), p. 483.

2. See, generally, W. E. H. Lecky, *The Rise and Influence of Rationalism in Europe* (New York: Braziller, 1955 ed.); Sir William Cecil Dampier(-Whetham), *A History of Science and Its Relations with Philosophy and Religion* (New York: Macmillan, rev. ed., 1943); Joseph Needham (ed.), *Science, Religion, and Reality* (New York: Braziller, 1955 ed.); Arthur Koestler, *The Sleepwalkers* (New York: Macmillan, 1959); J. H. Randall, Jr., *The Making of the Modern Mind* (New York: Houghton Mifflin, rev. ed., 1940), Bks. I and II.

3. See his *Pragmatism: A New Name for Some Old Ways of Thinking* (New York, 1907). Cf. the recent attempt at schematizing the polar traits of "tender" and "tough" mindedness by Heinz and Rowena Ansbacher, *The Individual Psychology of Alfred Adler* (New York: Basic Books, 1956), Introduction.

4. On "the quixotic attack of the young Hegel on the empirical method of science and his swift defeat at the hands of the scientists," see Michael Polanyi, *Personal Knowledge: Towards a Post-Critical Philosophy* (Chicago: University of Chicago Press, 1958), pp. 153 ff. On the career and ultimate fate of Goethe's *Naturphilosophie*, see Erich Heller, *The Disinherited Mind* (New York: Meridian Books, 1959), "Goethe and the Idea of Scientific Truth," pp. 3 ff. Heller observes: "Goethe's science has contributed nothing substantial to the scientific progress between his time and ours; . . . but he did, by his opposition to contemporary science, lay bare in his time, with remarkable precision, the very roots of that crisis and revolution in scientific method in which the twentieth-century scientist finds himself involved. In the history of science from Newton to Einstein, Goethe, the scientist, plays a Cinderella role, showing up the success and splendour of his rich relations, but also the potential *hubris* inherent in their pursuits." *Ibid.*, p. 8.

5. *Physics and Philosophy: The Revolution in Modern Science* (New York: Harper, 1958), p. 198. Another prominent quantum physicist has written of the same development: "This change in attitude may be characterized in a single phrase that also suggests its human content; it is a

turn from arrogance to humility." Pascual Jordan, *Science and the Course of History* (New Haven: Yale University Press, 1955), p. x. Cf. the strongly worded statement by C. F. von Weizsäcker, *The World View of Physics* (Chicago: University of Chicago Press, 1952), p. 36. See also the excellent analysis by F. Waismann, "The Decline and Fall of Causality," in *Turning Points in Physics* (New York: Harper Torchbooks, 1961), pp. 84–154.

6. Eddington has succinctly described the new direction taken by thermodynamics: "Alongside the super-intelligence imagined by Laplace for whom 'nothing would be uncertain' was placed an intelligence for whom nothing would be certain but some things would be exceedingly probable . . . the aim of science to approximate to this latter intelligence is by no means equivalent to Laplace's aim." *New Pathways In Science* (New York: Macmillan, 1935), pp. 76–77. Cf. Hans Reichenbach, *Atom and Cosmos: The World of Modern Physics* (New York: Macmillan, 1933), pp. 171 ff. See also Erwin Schrodinger, *What is Life? and Other Scientific Essays* (New York: Doubleday Anchor Books, 1956), pp. 236 ff.

7. Albert Einstein, *Out of My Later Years* (New York: Philosophical Library, 1950), p. 101. Cf. Werner Heisenberg, *The Physicist's Conception of Nature* (New York: Harcourt, Brace, 1958), p. 152.

8. Emery Neff, *The Poetry of History* (New York: Columbia University Press, 1947), p. 204. Prominent among the unconverted was the redoubtable Lord Kelvin, who confessed that he could understand nothing of which he could not make a mechanical model. Jeans has drily observed that Kelvin, "like many of the great scientists of the nineteenth century, stood high in the engineering profession: many others could have done so had they tried. It was the age of the engineer-scientist, whose primary ambition was to make mechanical models of the whole of nature." Sir James Jeans, *The Mysterious Universe* (New York: Dutton, 1958 [1930]), p. 31.

9. Einstein has pointed out that "it took physicists some decades to grasp the full significance of Maxwell's discovery, so bold was the leap that his genius forced upon the conceptions of his fellow-workers. . . . For several decades most physicists clung to the conviction that a mechanical substructure would be found for Maxwell's theory. But the unsatisfactory results of their efforts led to gradual acceptance of the new field concepts as irreducible fundamentals—in other words, physicists resigned themselves to giving up the idea of a mechanical foundation." *Out of My Later Years*, pp. 102–103.

10. Quoted in J. W. N. Sullivan, *Aspects of Science* (New York: 1925), p. 56.

11. The piscatorial analogy suggested here has been developed by Eddington, whose imaginary scientist-ichthyologist argues: "Anything uncatchable by my net is *ipso facto* outside the scope of ichthyological knowl-

edge, and is not part of the kingdom of fishes which has been defined as the theme of ichthyological knowledge. In short, what my net can't catch isn't fish. Or—to translate the analogy—'If you are not simply guessing, you are claiming a knowledge of the physical universe discovered in some other way than by the methods of physical science, and admittedly un-verifiable by such methods. You are a metaphysician. Bah!'" *The Philosophy of Physical Science* (Ann Arbor: University of Michigan Press, 1958 [1939]), p. 16.

12. Pertinent to this complacency is a comment of Whitehead: "Nothing is more curious than the self-satisfied dogmatism with which mankind at each period of its history cherishes the delusion of the finality of its existing modes of knowledge. . . . At this moment scientists and sceptics are the leading dogmatists. Advance in detail is admitted: fundamental novelty is barred. This dogmatic common sense is the death of philosophic adventure. The Universe is vast." Lucien Price (ed.), *Dialogues of Alfred North Whitehead* (New York: Mentor Books, 1956), p. 12.

13. An eminent physicist of the nineteenth century, Ernst Mach, gave dramatic testimony to this dogmatism in a letter to a colleague: "We can see that the physicists are on the surest road to becoming a church. . . . I hereby abandon the physical manner of thought, I will be no longer a regular physicist. I will renounce all scientific recognition; in short, the communion of the faithful I will decline with thanks. For dearer to me is freedom of thought." Quoted in Jerome Frank, *Fate and Freedom* (New York: Simon and Schuster, 1945), p. 104. Frank, in a somewhat overdrawn polemic against "our ascetic-fatalistic science," is accurate at least in his recognition that "many other men, not themselves scientists, came to maintain that this scientific determinism, so effective in revolutionizing industry, was the whole truth. Thus modern scientific fatalism became a faith, a religion." *Ibid.*, pp. 104, 87–105 *passim*.

14. J. W. N. Sullivan, *The Limitations of Science* (New York: Mentor Books, 1949 [1933]), p. 140.

15. "But with all of this, and with varying degrees of agreement and reservation, there was the belief that in the end all nature would be reduced to physics, to the giant machine. Despite all the richness of what men have learned about the world of nature, of matter and of space, of change and of life, we carry with us today an image of the giant machine as a sign of what the objective world is really like." J. Robert Oppenheimer, *Science and the Common Understanding* (New York: Simon and Schuster, 1954), pp. 14–15.

16. *The Limitations of Science*, pp. 140–141. (Emphasis added.)

17. Heisenberg, *The Physicist's Conception of Nature*, p. 152. "Hertz points out that propositions in physics have neither the task nor the capacity of revealing the inherent essence of natural phenomena. He concludes that physical determinations are only pictures, on whose corre-

spondence with natural objects we can make but the single assertion, viz., whether or not the *logically* derivable consequences of our pictures correspond with the empirically observed consequences of the phenomena for which we have designed our picture." *Ibid.*, pp. 152–153.

18. Dampier has summarized his contribution as follows: ". . . Mach pointed out that science does but construct a model of what our senses tell us about Nature, and that mechanics, far from being necessarily the ultimate truth about Nature as some believe it to be, is but one aspect from which that model may be regarded." *A History of Science*, p. 317. The stark contrast of this view with the prevailing dogma is aptly described by the same historian: "But few nineteenth century men of science were interested in philosophy, even in that of Mach. Most of them assumed that they themselves were dealing with realities, and that the main lines of possible scientific enquiry had been laid down once for all. It seemed that all that remained for the physicist to do was to make measurements to an increasing order of accuracy, and invent an intelligible mechanism which would explain the nature of the luminiferous aether." *Ibid.*, p. xix.

19. For example, in the preface to the second edition of his *Grammar*, Pearson wrote: "Step by step men of science are coming to recognize that mechanism is not at the bottom of phenomena, but is only the conceptual shorthand by aid of which they can briefly describe and resume phenomena. That all science is description and not explanation, that the mystery of change in the inorganic world is just as great and just as omnipresent as in the organic world, are statements which will appear as platitudes to the next generation." *The Grammar of Science* (New York: Meridian Books, 1957 ed.), p. xiv. For convenient summaries of Pearson's "phenomenalism," cf. *ibid.*, pp. xi, 177, 218, *passim*.

20. *Out of My Later Years*, p. 102.

21. Lincoln Barnett, *The Universe and Dr. Einstein* (New York: Mentor Books, rev. ed., 1958), p. 41.

22. *Ibid.*

23. *The Limitations of Science*, p. 139.

24. *The Mysterious Universe*, p. 101. "We must always remember that the existence of the ether is only an hypothesis, introduced into science by physicists who, taking it for granted that everything must admit of a mechanical explanation, argued that there must be a mechanical medium to transmit waves of light, and all other electrical and magnetic phenomena. To justify their belief, they had to show that a system of pushes, pulls and twists could be devised in the ether to transmit all the phenomena of nature through space and deliver them up at the far end exactly as they are observed—much as a system of bell-wires transmits mechanical force from a bell-pull to a bell." *Ibid.*, p. 112.

25. Quoted in Jeans, *ibid.*, p. 101.

26. Kelvin is quoted as remarking in a lecture: "This thing we call the luminiferous ether . . . is the only substance we are confident of in dynamics. One thing we are sure of, and that is the reality and substantiality of the luminiferous ether." E. N. da C. Andrade, *An Approach to Modern Physics* (New York: Doubleday Anchor Books, 1957), p. 242.

27. Sullivan, *The Limitations of Science*, p. 140.

28. Cf. Heisenberg, *The Physicist's Conception of Nature*, p. 15.

29. Jeans, *The Mysterious Universe*, p. 119.

30. Cf. Louis de Broglie, *The Revolution in Physics* (New York: Noonday Press, 1953), pp. 100–101, 79–94 *passim*. For a more aggressive (if not defensive-aggressive) statement of the consistency of classical and relativistic theory, see Max Planck, *The New Science* (New York: Meridian Books, 1959), pp. 163–165.

31. The invention of the non-Euclidean geometries has been described by Sullivan as "one of the most remarkable feats in the intellectual history of man. For two thousand years Euclid's axioms had reigned unchallenged. That they were 'necessary truths,' true for angels as well as men, true even for God Himself, was admitted by all the philosophers. Merely to wonder whether these truths could be transcended, merely to wonder whether there was a world of ideas outside them, was an effort of extraordinary imaginative daring." *The Limitations of Science*, p. 18. For an excellent brief account of the development of the non-Euclidean geometries and their physical applications, see Sir Edmund Whittaker, *From Euclid to Eddington* (New York: Dover Publications, 1958), pp. 31 ff. *passim*.

32. *Science and the Modern World*, pp. 17–18.

33. He continues: "Yet in a completely objective survey of the situation, the outstanding fact would seem to be that mechanics has already shot its bolt and has failed dismally, on both the scientific and philosophical side. If anything is destined to replace mathematics, there would seem to be specially long odds against it being mechanics." *The Mysterious Universe*, p. 179.

34. Cf. the comments of de Broglie under the heading, "Classical Mechanics and Physics are Approximations," in *The Revolution in Physics*, pp. 20 ff. See also the suggestive discussion by Max Born, "Is Classical Mechanics in Fact Deterministic?" in *Physics in My Generation* (London and New York: Pergamon Press, 1956), pp. 164–170.

35. Sullivan, *The Limitations of Science*, p. 56.

36. It was Boyle who pointed out "that the primary qualities or geometrical concepts in terms of which mathematical physics organised and interpreted experience were no less mental than the secondary qualities, and that if either group had claim to reality then both had equal claims."

A. C. Crombie, *Medieval and Early Modern Science*, Vol. II, p. 310. Subsequently Leibniz and, more consistently, Berkeley were to advance similar criticisms.

37. Barnett, *The Universe and Dr. Einstein*, p. 19. Cf. Sullivan, *The Limitations of Science*, p. 55: "Nature, it appears, knows nothing of the distinction we make between space and time. The distinction we make is, ultimately, a psychological peculiarity of ours."

38. *The Philosophy of Physical Science*, p. 57. (Emphasis in the original.) Eddington did not of course mean that all scientific knowledge is wholly subjective; he did however develop an epistemological theory which he called "Selective Subjectivism," in which the subjective elements strongly predominated over the objective content. (*Ibid.*, chap. 2.)

39. *The Mysterious Universe*, p. 181. Jeans went on, as is well known, to argue that since the universe has the appearance of a great thought, it must be a thought in the mind of a Great Mathematician. Distaste for this spiritual culmination of his ideas has led, except in theological circles, to a rather summary disregard or dismissal of the rest of Jeans's argument. But that argument, aside from its substantial agreement with Eddington's, is also largely congenial with the scientific philosophy espoused in subsequent years by the leading quantum physicists. (See below.)

40. Something of the scope of the quantum revolution is indicated by the comment of James B. Conant: "All this is quite different from a so-called revolutionary discovery like the discovery of radioactivity; it is more closely akin to the formulation of such epoch-making new concepts as those embodied in Newtonian mechanics or Darwin's theory of evolution. Yet some would probably maintain that the new physics is more of a revolution, represents more of a break with the past than has the introduction of any new theory in science since 1600." *Modern Science and Modern Man* (Garden City, N. Y.: Doubleday Anchor Books, 1954), pp. 66–67.

41. See Max Planck, *The New Science*, pp. 18 ff.; Bertrand Russell, *Human Knowledge: Its Scope and Limits* (New York: Simon and Schuster, 1948), pp. 22 ff.

42. See A. d'Abro, *The Rise of the New Physics* (New York: Dover Publications, 1951), Vol. II, p. 944. Cf. the statement of Jeans: "Before the quantum theory appeared, the principle of the uniformity of nature—that like causes produce like effects—had been accepted as a universal and indisputable fact of science. As soon as the atomicity of radiation became established, this principle had to be abandoned." Sir James Jeans, *Physics and Philosophy* (New York: Macmillan, 1943), p. 140. Cf. Stephen Toulmin, *The Philosophy of Science* (London: Hutchinson, 1953), pp. 148, 151.

43. Cf. Whittaker, *From Euclid to Eddington*, pp. 136 ff.; Barnett, *The Universe and Dr. Einstein*, pp. 24 ff.

44. De Broglie, *The Revolution in Physics*, p. 14. In a memorial tribute to Planck, Einstein further underlined the importance of his theory: "This discovery became the basis of all twentieth-century research in physics and has shattered the whole framework of classical mechanics and electro-dynamics and set science a fresh task: that of finding a new conceptual basis for all physics." "Max Planck in Memoriam," in Einstein, *Out of My Later Years*, pp. 229–230.

45. In testimony before the Joint Congressional Atomic Energy Subcommittee, as reported in the *New York Times* of July 15, 1959, Dr. Leland J. Haworth, director of the Brookhaven National Laboratory, stated that "as man explores at shorter and shorter distances in this sub-microscopic world . . . it may turn out that he has been wrong in thinking that space is continuous, like time. It may be found, he said, that there is a length, or what scientists call a 'quantum' in space that represents a minimum distance in space. Such a discovery, Haworth said, would force 'a radical revision of our concept of space itself' and 'raise questions of great philosophic and scientific interest.' " This possibility was, however, anticipated at least as early as the 1930's: cf. Sullivan, *The Limitations of Science*, p. 63.

46. Abro, *Rise of the New Physics*, Vol. II, p. 446.

47. De Broglie, *The Revolution in Physics*, p. 126.

48. L. Susan Stebbing, *Philosophy and the Physicists* (New York: Dover Publications, 1958 [1937]), p. 177: "It follows, then, from Bohr's theory that the behavior of an individual atom is not determined in accordance with any known laws. . . . It is important to notice that this theory involves a complete break with the deterministic scheme of law of classical physics. . . . Subsequent progress has been in the direction of elaborating theories that depart more and more from classical conceptions. But even in its first form Bohr's theory necessitated a break with classical theories and provided the first set of arguments in favour of indeterminism." Cf. Max Born, *Physics in My Generation*, pp. 125–126.

49. The role and stature of Bohr is indicated by Max Born's description of him as "the founder of modern atomic theory, and the deepest thinker in physical science." *Physics in My Generation*, p. 42. Cf. the comment of Weizsäcker: "In this respect the present essays do not pretend to be more than a commentary on Bohr's views. It is Bohr who has influenced the philosophical ways of thought of all younger atomic physicists in the most profound manner, and I want to use this occasion for expressing once more the admiration and gratitude we all feel for his influence." *The World View of Physics*, p. 6.

50. Cf. Abro, *Rise of the New Physics*, Vol. II, pp. 944–949

passim; Sullivan, *The Limitations of Science*, p. 69; Born, *Physics in My Generation*, p. 45.

51. De Broglie, *The Revolution in Physics*, p. 212. Cf. de Broglie, *Physics and Microphysics* (New York: Harper Torchbooks, 1960), pp. 114 ff.

52. Born, *Physics in My Generation*, p. 124. Cf. Abro, *Rise of the New Physics*, Vol. II, pp. 947–948; Niels Bohr, *Atomic Physics and Human Knowledge* (New York: John Wiley and Sons, 1958), p. 19.

53. De Broglie has stated the reasoning as follows: "The product of the uncertainty of a coordinate by the uncertainty of the corresponding component of momentum is always at least of the order of magnitude of Planck's constant h." *The Revolution in Physics*, p. 207.

54. Cf. J. Bronowski, *The Common Sense of Science* (Cambridge: Harvard University Press, 1955), p. 72; Abro, *Rise of the New Physics*, Vol. II, pp. 948–949. Bridgman states the case with uncompromising vigor: "The same situation confronts the physicist everywhere; whenever he penetrates to the atomic or electronic level in his analysis, he finds things acting in a way for which he can assign no cause, for which he never can assign a cause, and for which the concept of cause has no meaning, if Heisenberg's principle is right. This means nothing more nor less than that the law of cause and effect must be given up." P. W. Bridgman, *Reflections of a Physicist* (New York: Philosophical Library, 1950), p. 93.

55. Weizsäcker, *The World View of Physics*, p. 47.

56. Oppenheimer, *Science and the Common Understanding*, p. 62.

57. *The Common Sense of Science*, p. 68. One of the most cogent presentations of the "a-causal" statistical viewpoint is that of Erwin Schrödinger, "What is a Law of Nature?" in his *Science and the Human Temperament* (New York: Norton, 1935), subsequently republished with additions as *Science Theory and Man* (New York: Dover, 1957). The following passage is especially noteworthy: "Whence arises the widespread belief that the behavior of molecules is determined by absolute causality, whence the conviction that the contrary is *unthinkable*? Simply from the *custom*, inherited through thousands of years, of *thinking causally*, which makes the idea of undetermined events, of absolute, primary casualness, seem complete nonsense, a logical absurdity. But from what source was this habit of causal thinking derived? Why, from observing for hundreds and thousands of years precisely *those regularities* in the natural course of events which, in the light of our present knowledge, are most certainly *not governed by causality;* or at least not so governed essentially, since we now know them to be *statistically* regulated phenomena. Therewith this traditional habit of thinking loses its rational foundation." *Science Theory and Man*, pp. 143–144 (emphasis in original). (Note the correction, in the

second quoted sentence above, of typographical errors in both editions of the book which unintentionally reverse the meaning of the passage.)

58. *The Common Sense of Science*, p. 68. The psychological, if not the essential, difference between the absolute certainty of classical determinism and the contingency introduced by quantum physics was graphically expressed in an address delivered by William James in 1884, long before the advent of the quantum theory: "The stronghold of the deterministic sentiment is the antipathy to the idea of chance. As soon as we begin to talk indeterminism to our friends, we find a number of them shaking their heads. This notion of alternative possibility, they say, this admission that any one of several things may come to pass, is after all, only a roundabout name for chance: and chance is something the notion of which no sane mind can for an instant tolerate in the world. What is it, they ask, but barefaced crazy unreason, the negation of intelligibility and law? And if the slightest particle of it exist anywhere, what is to prevent the whole fabric from falling together, the stars from going out, and chaos from recommencing her topsy-turvy reign?" "An Address to Harvard Divinity Students," reprinted in Albert William Levi, *Varieties of Experience* (New York: Ronald Press, 1957), p. 205.

59. "The world from the standpoint of mechanics is an automaton, without any freedom, determined from the beginning. I never liked this extreme determinism, and I am glad that modern physics has abandoned it." *Physics in My Generation*, p. 45. Cf. Weizsäcker, *The World View of Physics*, p. 36; Heisenberg, *Physics and Philosophy*, pp. 197–198; Bohr, *Atomic Physics and Human Knowledge*, pp. 94, 96 *passim;* Andrade, *An Approach to Modern Physics*, pp. 244, 250, 257; Eddington, *New Pathways in Science*, pp. 72 ff.; Dampier, *A History of Science*, pp. 472 ff.

60. Einstein agreed that "Heisenberg has convincingly shown, from an empirical point of view, any decision as to a rigorously deterministic structure of nature is definitely ruled out, because of the atomistic structure of our experimental apparatus. Thus it is probably out of the question that any future knowledge can compel physics again to relinquish our present statistical theoretical foundation in favor of a deterministic one which would deal directly with physical reality." But he also wrote: "Some physicists, among them myself, can not believe that we must abandon, actually and forever, the idea of direct representation of physical reality in space and time; or that we must accept the view that events in nature are analogous to a game of chance." *Out of My Later Years*, pp. 109–110. For discussion of Einstein's attitude, see Born, *Physics in My Generation*, pp. 201 ff. Planck was rather less equivocal: ". . . I must definitely declare my own belief that the assumption of a strict dynamic causality is to be preferred simply because the idea of a dynamically law-governed universe is of wider and deeper application than the merely statistical idea . . ." *The New Science*, p. 58.

61. Quoted in Stebbing, *Philosophy and the Physicists*, p. 185. Eddington wrote elsewhere: "Ten years ago practically every physicist of repute was, or believed himself to be, a determinist, at any rate so far as inorganic phenomena are concerned. . . . The withdrawal of physical science from an attitude it had adopted consistently for more than 200 years is not to be treated lightly; and it provokes a reconsideration of our views as to one of the most perplexing problems of our existence." *New Pathways in Science*, pp. 72–73.

62. Sullivan, *The Limitations of Science*, p. 70.

63. *An Approach to Modern Physics*, p. 255.

64. Cf. Born, *Physics in My Generation*, pp. 39–40, 124–125; de Broglie, *Physics and Microphysics*, pp. 114 ff., 118, 131–132.

65. *Science and the Common Understanding*, p. 13.

66. *Physics in My Generation*, p. 105. Cf. Bohr, *Atomic Physics and Human Knowledge*, pp. 20, 62, *passim*. Cf. also Victor F. Lenzen, *Causality in Natural Science* (Springfield, Ill.: Thomas, 1954), p. 109.

67. Bronowski asserts that "In the physical and in the logical worlds, what we have really seen happen is the breakdown of the plain model of a world outside ourselves where we simply look on and observe. . . . For Relativity derives essentially from the philosophic analysis which insists that there is not a fact and an observer, but a joining of the two in an observation . . . that event and observer are not separable." *The Common Sense of Science*, p. 77. Cf. David Lindsay Watson, *The Study of Human Nature* (Yellow Springs, Ohio: Antioch Press, 1953), p. 63. "Yet we do not need to wait for such a synthesis to be able to grasp the kinship of indeterminacy and relativity for all knowledge. The crux of the matter is this: *the existence of such a linkage* (between the knower and the thing known) *determines in part the nature* of the event observed, indeed, often largely controls it." (Emphasis in original.)

68. "Intimately connected with this duality is the polarity sub-jective-objective. For if an experiment must be set up in a definite way to investigate one or the other of a conjugate pair of quantities, it is impossible to obtain information of the system considered as such; the observer has to decide beforehand which kind of answer he wants to obtain. Thus sub-jective decisions are inseparably mixed with objective observations." Born, *Physics in My Generation*, p. 128.

69. *The World View of Physics*, pp. 178, 200. For a systematic development of the thesis of personal responsibility and commitment on the part of the scientist, see Polanyi, *Personal Knowledge* (cited above, Note 4). Cf. W. I. B. Beveridge, *The Art of Scientific Investigation* (New York: Norton, n.d.); C. Judson Herrick, *The Evolution of Human Nature* (Austin: University of Texas Press, 1956), p. 61, *passim.*; Max Otto, *Science and the Moral Life* (New York: Mentor Books, 1949), pp. 99 ff.

70. Heisenberg, *The Physicist's Conception of Nature*, p. 24. (Emphasis in original.) De Broglie writes similarly: "Science thus loses a part of its objective character; no more is it the passive contemplation of a fixed universe; it becomes a hand-to-hand struggle where the scientist succeeds in snatching from the physical world, which he would like to understand, certain information, always partial, which would allow him to make predictions that are incomplete, and in general, only *probable.*" *Physics and Microphysics*, p. 131.

71. Bridgman, *Reflections of a Physicist*, p. 96 (emphasis added). Cf. Bridgman, *The Nature of Physical Theory* (New York: Dover, n.d.), pp. 121 ff. *passim.*

72. See John Dewey and Arthur F. Bentley, *Knowing and the Known* (Boston: Beacon Press, 1949). For earlier formulations of this viewpoint in terms of interaction, see John Dewey, *The Quest for Certainty* (New York: Minton, Balch, 1929), and *Experience and Education* (New York: Macmillan, 1938); Arthur F. Bentley, *Behavior, Knowledge, Fact* (Bloomington, Ind.: Principia Press, 1935). The development and application of this "contextualist" approach may be variously traced in G. H. Mead, *Mind, Self and Society* (Chicago: University of Chicago Press, 1934); M. P. Follett, *Creative Experience* (West Ridge, N.H.: Smith, 1951); W. H. Ittelson and Hadley Cantril, *Perception: A Transactional Approach* (Garden City: Doubleday, 1954). Cf. Wallace A. MacDonald, "The Social Scientist as Actor." *Berkeley Publications in Society and Institutions* (Berkeley, 1957), Vol. III, pp. 1–12.

73. Quoted in Heisenberg, *The Physicist's Conception of Nature*, p. 153.

74. For significant demurrers to this conventional view with respect to living organisms, both macroscopic and microscopic, see the discussion in Chap. 5, below, "The Microphysics of Life."

75. "Although, then, there is no perceptible change in our practical activity [as a result of quantum mechanics], the revolution in our theoretical knowledge is all the more profound. What happens is not predetermined in all details, as determinism, distorting world history into the mechanical performance of a clock movement, maintains; the course of all events is much more like a continual game of dice, so that each separate step corresponds to a new throw. The decision, as between a causal and a statistical view of the world, has fallen in favour of statistics." Hans Reichenbach, *Atom and Cosmos*, pp. 278–279.

76. *The Revolution in Physics*, p. 22.

77. Cf. Abro, *Rise of the New Physics*, p. 949. Another physicist has described the situation as follows: "But if the old notions of classical mechanics do not apply in the atomic world, they also cannot be *absolutely* correct in regard to the motion of larger material bodies. Thus we are led

to the conclusion that the principles underlying classical mechanics must be considered only as very good approximations to the 'real thing,' approximations that fail badly as soon as we try to apply them to systems more delicate than those for which they were originally intended." George Gamow, *One Two Three . . . Infinity* (New York: Mentor Books, 1953), p. 141.

78. Ludwig Lewisohn has somewhere characterized this attitude as "the intellectual integrity of the scientific mechanist who has made a religion of an exactness which does not exist." Cf. William James's description of the determinist psychology, cited above, Note 58.

79. *New Pathways in Science*, p. 80.

80. *Ibid.* Cf. J. von Neumann, *Mathematische Grundlagen der Quantenmechanik* (Berlin: Spring, 1932), and the comment of de Broglie, *The Revolution in Physics*, p. 217n.

81. *The Common Sense of Science*, p. 68.

82. Heisenberg, *The Physicist's Conception of Nature*, p. 42; Heisenberg, *Physics and Philosophy*, p. 154. By far the strongest, not to say the most radical, argument for the direct influence of quantum laws upon living organisms at all levels of complexity is that of Pascual Jordan, *Physics of the Twentieth Century* (New York: Philosophical Library, 1944), pp. 150 ff. His views are discussed below, in connection with complementarity.

83. De Broglie, *The Revolution in Physics*, p. 22. (These words were written in 1936.)

84. See the similar statement by Max Born, *Physics in My Generation*, p. 107.

85. Cf. de Broglie, *The Revolution in Physics*, pp. 157 ff., 190 ff.; Abro, *Rise of the New Physics*, pp. 637 ff.; Andrade, *An Approach to Modern Physics*, pp. 138 ff.

86. Cf. Sullivan, *The Limitations of Science*, pp. 66 ff.

87. The concept was apparently first enunciated in a 1927 article, "The Quantum Postulate and the Recent Development of Atomic Theory," reprinted in Niels Bohr, *Atomic Theory and the Description of Nature* (Cambridge, England: Cambridge University Press, 1934), pp. 52 ff. Cf. *ibid.*, "Introductory Survey," pp. 10 ff. See also Bohr, "Causality and Complementarity," *Philosophy of Science*, Vol. IV (1937), pp. 189 ff.; "On the Notions of Causality and Complementarity." *Dialectica*, Vol. II (1948), pp. 312 ff.

88. De Broglie, *The Revolution in Physics*, p. 218. Cf. Jordan, *Physics of the Twentieth Century*, p. 132.

89. Max Born has intimated that this is the original and correct

application of the complementarity principle, and prefers to regard the particle-wave relationship as a "duality." See his *Natural Philosophy of Cause and Chance* (Oxford: Clarendon Press, 1949), p. 105.

90. Louis de Broglie, *Matter and Light: The New Physics* (New York: Horton, 1939), p. 279. The phrase "tolerance of ambiguity" was first given currency by the late Else Frenkel-Brunswik, one of the team of social scientists who produced *The Authoritarian Personality* (New York: Harper, 1950). The psychological category "tolerance-intolerance toward ambiguity" served as an index of the degree of "authoritarianism," or alternately of "liberalism," manifested by the subjects tested in this research. Cf. Frenkel-Brunswik, "Tolerance Toward Ambiguity as a Personality Variable," (Abstract) *American Psychologist*, Vol. III (1948), pp. 268 ff. It might be interesting to speculate on the degree to which the modern quest for certainty among scientists and social scientists, and more especially the Laplacean dogma of absolute objective determinism, reflects an "authoritarian" intolerance of ambiguity. Lest this suggestion appear merely jocular, it is pertinent to note that the Soviet Union has, at least until recently, prohibited any deviation among its scientists from strict and thorough determinism, and that, in the words of one authority, "Special attacks were leveled against scientists who proclaimed the basic acausality of single atomic events (and thus countenanced so-called statistical determinacy); recognized the complementary properties of atomic systems; or offered 'subjectivistic,' 'formalistic' or 'operationalist' interpretations of the theory of relativity." Alexander Vucinich, "The Ethics of Soviet Science," *Pacific Spectator*, Vol. IX (1955), p. 333.

91. *Matter and Light*, p. 280. Cf. de Broglie, *The Revolution in Physics*, pp. 218–219.

92. Born, *Physics in My Generation*, p. viii.

93. J. Robert Oppenheimer, *The Open Mind* (New York: Simon and Schuster, 1955), p. 82: "You know that when a student of physics makes his first acquaintance with the theory of atomic structure and of quanta, he must come to understand the rather deep and subtle notion which has turned out to be the clue to unraveling that whole domain of physical experience. This is the notion of complementarity which recognizes that various ways of talking about physical experience may each have validity, and may each be necessary for the adequate description of the physical world, and may yet stand in a mutually exclusive relationship to each other, so that to a situation to which one applies, there may be no consistent possibility of applying the other."

94. J. L. Destouches, "Quelques Aspects Theoriques de la Notion de Complémentarité," *Dialectica*, Vol. II (1948), pp. 351 ff. Lenzen has summarized this presentation as follows: "M. Destouches divides physical theories into two classes: 1) Theories for which there exists a state quantity and which are objectivist, for example, classical theories; 2) Theories for

which there exists no state quantity and which are subjectivist, for example, wave mechanics. . . . Objectivist theories are characterized by causality, subjectivist theories by complementarity. An objectivist theory is deterministic, a subjectivist theory is indeterministic." Lenzen, *Causality in Natural Science*, pp. 105–106.

95. See Paulette Destouches-Fevrier, "Manifestations et Sens de la Notion de Complémentarité," *Dialectics*, Vol. II (1948), pp. 351 ff. Cf. Lenzen, *Causality in Natural Science*, pp. 107–109; G. Birkhoff and J. von Neumann, "The Logic of Quantum Mechanics," *Annals of Mathematics*, Vol. XXXVII (1936), pp. 823 ff.; Hans Reichenbach, *The Philosophic Foundations of Quantum Mechanics* (Berkeley: University of California Press, 1944), pp. 144 ff.; Heisenberg, *Physics and Philosophy*, pp. 181 ff.; Gustav Bergmann, "The Logic of Quanta," in Herbert Feigl and May Brodbeck (eds.), *Readings in the Philosophy of Science* (New York: Appleton-Century-Crofts, 1953), pp. 475–509.

96. *Physics in My Generation*, p. 107. Cf. Jordan, *Physics of the Twentieth Century*, p. 131: "This idea of complementarity must be viewed as the most significant result for philosophy that has crystallized out of modern physics. . . . it appears justified to believe that it may further become of epoch-making importance in other realms of natural science."

97. Cf., *inter alia*, Bohr, *Atomic Physics and Human Knowledge*, pp. 27, 62 ff., 96, *passim*; Heisenberg, *Physics and Philosophy*, pp. 106, 154–155, 199; Jordan, *Physics of the Twentieth Century*, pp. 150 ff.; Bronowski, *The Common Sense of Science*, pp. 46, 61 ff., 76, *passim*; Polanyi, *Personal Knowledge*, pp. 336 ff.; Born, *Physics in My Generation*, p. 52; Weizsäcker, *The History of Nature* (Chicago: University of Chicago Press, 1949), pp. 122 ff., 172 ff.; Bridgman, *Reflections of a Physicist*, p. 312.

98. See his *Atomic Physics and Human Knowledge*, *passim*; and his *Atomic Theory and the Description of Nature*, pp. 22–24.

99. See Jordan, *Physics of the Twentieth Century*, pp. x–xi, 150 ff.; Heisenberg, *Physics and Philosophy*, esp. pp. 103 f., 106 ff.; de Broglie, *Matter and Light*, pp. 253, 261, 281 ff.; Sullivan, *Limitations of Science*, pp. 107 f., 125 ff. See also the highly individual contribution of Schrödinger, *What is Life?*, which defies classification.

100. Bohr, *Atomic Physics and Human Knowledge*, pp. 21 f., 27. The significance of the "understanding" approach, both for psychology and political science, is considered in Chapters VI and VII, below. Representative expressions of this viewpoint are Gordon W. Allport, *Becoming* (New Haven: Yale University Press, 1955), and A. H. Maslow, *Motivation and Personality* (New York: Harper, 1954). Cf. Theodore Abel, "The Operation Called *Verstehen*," in Feigl and Brodbeck (eds.), *Readings in Philosophy of Science*, pp. 677–687; Alan Gewirth, "Subjectivism and Objectivism in the Social Sciences," *Philosophy of Science*, Vol. XXI (1954), pp. 175–183.

101. "But the fact that science does not know the 'Thou' leads us to an entirely different question: not whether one *can* experiment with man, but whether one wants to or may experiment with him. What do I do to my fellow man in treating him in my thoughts or actions as a mere object? In this question seems to focus everything that has rightly been advanced against the application of natural science to man." Weizsäcker, *The World View of Physics,* p. 206.

102. See Bohr, *Atomic Theory and the Description of Nature,* pp. 22 ff. The crux of the statement is that "there is set a fundamental limit to the analysis of the phenomena of life in terms of physical concepts, since the interference necessitated by an observation which would be as complete as possible from the point of view of the atomic theory would cause the death of the organism. In other words: *the strict application of those concepts which are adapted to our description of inanimate nature might stand in a relationship of exclusion to the consideration of the laws of the phenomena of life.*" (Emphasis in the original.) This would seem to be a very literal translation of the moral admonition well expressed by Tillich: "Man resists objectification, and if his resistance to it is broken, man himself is broken." (Paul Tillich, *Systematic Theology* [Chicago: University of Chicago Press, 1951], Vol. I, p. 98.) An obvious application may be seen in medicine, where, as Muller has commented, the mechanistic bias has resulted in great harm if not actual death, especially in the treatment of mental disease and in resistance to such therapeutic innovations as psychoanalysis. Herbert J. Muller, *Science and Criticism* (New York: Braziller, 1956 ed.), p. 108n.

103. See, for example, B. F. Skinner, "The Control of Human Behavior," *Transactions of the New York Academy of Sciences,* Ser. II, Vol. XVII, No. 7, pp. 547–551; S. S. Stevens, "Measurement and Man," *Science,* Vol. CXXVII (1958), pp. 386 ff.; Adolph Grunbaum, "Causality and the Science of Human Behavior," *American Scientist,* Vol. XL (1952), pp. 665–676.

104. Michael Polanyi, *The Logic of Liberty* (Chicago: University of Chicago Press, 1951), p. 22. For statements similar to that of Polanyi, cf. Kenneth Burke, *Permanence and Change* (Los Altos, California: Hermes, 1954 ed.), pp. 218–219; Donald Snygg and A. W. Combs, *Individual Behavior* (New York: Harper, 1949), p. 24n.

105. See Note 104, above. Bohr himself has recognized the philosophical antiquity of his principle; see *Atomic Physics and Human Knowledge,* pp. 20, 62. More than one observer has pointed to the similarities between complementarity and dialectical philosophies; e.g., J. L. Destouches (article cited above, Note 94), and Max Born, *Physics in My Generation,* p. 107. Another apparently equivalent philosophical doctrine is Northrop's concept of the "two-termed relation of epistemic correlation" uniting the "aesthetic" and "theoretic" components of reality, of which he writes:

"Both components are equally real and primary, and hence good, the one being the complement of the other." F. S. C. Northrop, *The Meeting of East and West* (New York: Macmillan, 1949), p. 450. Cf. Northrop, "Complementary Emphases of Eastern Intuitive and Western Scientific Philosophy," in C. A. Moore (ed.), *Philosophy: East and West* (Princeton: Princeton University Press, 1944).

106. On this point, see especially F. A. Hayek, *The Sensory Order: An Inquiry Into the Foundations of Theoretical Psychology* (London: Routledge and Kegan Paul, 1952). "The proposition which we shall attempt to establish is that any apparatus of classification must possess a structure of a higher degree of complexity than is possessed by the objects which it classifies; and that, therefore, the capacity of any explaining agent must be limited to objects with a structure possessing a degree of complexity lower than its own. If this is correct, it means that no explaining agent can ever explain objects of its own kind, or of its own degree of complexity, and, therefore, that the human brain can never fully explain its own operations." *Ibid.*, p. 185. Cf. Bridgman, *The Nature of Physical Theory*, pp. 122 f.

107. *Matter and Light*, p. 280. Cf. the very similar statement by Waisman, "The Decline and Fall of Causality" (cited above, Note 5), p. 113.

108. *Ibid.*, p. 281. In this connection de Broglie has written elsewhere: "We also could . . . examine whether all 'idealizations' are not that much less applicable to reality when they become more complete and, although we have little inclination to be paradoxical, we could hold, contrary to Descartes, that nothing is more misleading than a clear and distinct idea." *The Revolution in Physics*, p. 219. Cf. P. W. Bridgman, *The Way Things Are* (New York: Viking Press Compass Books, 1961), pp. 9 f.

109. *Matter and Light*, p. 282. Cf. Heisenberg, *Physics and Philosophy*, pp. 201–202: "The skepticism [of modern physics] against precise scientific concepts does not mean that there should be a definite limitation for the application of rational thinking. On the contrary, one may say that the human ability to understand may be in a certain sense unlimited. But the existing scientific concepts cover always only a very limited part of reality, and the other part that has not yet been understood is infinite."

110. Arthur H. Compton, *The Freedom of Man* (New Haven: Yale University Press, 1935). Cf., for a later statement, Arthur H. Compton, *The Human Meaning of Science* (Chapel Hill: University of North Carolina Press, 1940), esp. pp. viii–ix, 41–42.

111. *Atom and Cosmos*, pp. 279–280. (Emphasis added.) It is equally appropriate, in closing this chapter, to note the concluding words of Ernst Cassirer's *Determinism and Indeterminism in Modern Physics* (New Haven: Yale University Press, 1956), pp. 212–213: "What modern physics has taught us is the fact that the change of standpoint which we

have to make whenever we move from one dimension of meaning to another, whenever we exchange the world of science for that of ethics, art, etc., is not confined to this type of transition alone. The manifold of perspectives which opens up before us has its counterpart within the scientific realm itself. . . . *When, even in science, such a superposition of dissimilar aspects is necessary, it will be the more easily understandable that we shall meet such a superposition again as soon as we go outside its realm*—as soon as we seek to realize the full concept of reality, which requires the cooperation of all functions of the spirit and can only be reached through all of them together." (Emphasis added.)

CHAPTER V *The Ambiguity of Life*

1. Quoted in W. I. B. Beveridge, *The Art of Scientific Investigation* (New York: Norton, n.d.), p. 61.

2. Hans Reichenbach, *Atom and Cosmos: The World of Modern Physics* (New York: Macmillan, 1933), p. 270.

3. Quoted in Herbert J. Muller, *Science and Criticism* (New York: Braziller, 1956 ed.), p. 106.

4. Reichenbach, *Atom and Cosmos*, p. 270.

5. "Materialism, dialectical or otherwise, is a form of faith, founded on predilection and belief, which has an appeal for certain minds, but it certainly has no support from the findings or the founders of modern science." E. N. da C. Andrade, *An Approach to Modern Physics* (New York: Doubleday, 1956), p. 259.

6. Max Born, *Physics in My Generation* (London: Pergamon Press, 1956), p. 52.

7. The future history of biology, von Bertalanffy has suggested, may well contain a chapter entitled "The Struggle for the Concept of Organism in the Early Twentieth Century," recounting among other things "how the concept of organism was first taken seriously not by biologists but by some philosophers and mathematical physicists." Ludwig von Bertalanffy, *Problems of Life* (New York: Harper Torchbooks, 1960), p. 197.

8. *Ibid.*, p. 178. Cf. Louis de Broglie, *Matter and Light: The New Physics* (New York: Norton, 1939), pp. 279 ff.

9. A. N. Whitehead, *Science and the Modern World* (New York: Mentor Books, 1948), p. 105, *passim.* Cf. Whitehead, *Process and Reality* (New York: Macmillan, 1929).

10. Lancelot Law Whyte, *The Next Development in Man* (New York: Mentor Books, 1950), pp. 10–11.

11. Panpsychism, as a concept of the modern world claiming scientific status, may be traced backward at least to the belief of Giordano Bruno that "all reality is one in substance, one in cause, and one in origin; . . . every particle of reality is composed inseparably of the physical and the psychical." Quoted in G. Murray McKinley, *Evolution: The Ages and Tomorrow* (New York: Ronald Press, 1956), p. 4. Cf. O. L. Reiser, *Philosophy and the Concepts of Modern Science* (New York: Macmillan, 1935); A. C. Garnett, "Scientific Method and the Concept of Emergence," *Journal of Philosophy*, Vol. XXXIX (1942), pp. 477–486.

12. *Problems of Life*, p. 180.

13. William G. Pollard, "The Significance of Complementarity for the Life Sciences," *American Journal of Physics*, Vol. XX (1952), p. 283. Cf., for a more complete statement, Pollard, *Chance and Providence* (New York: Scribner's, 1958), esp. pp. 138–152.

14. "The Significance of Complementarity for the Life Sciences," pp. 285, 286, *passim*. The suggestion that it may become the turn of biology to influence the concepts of physical science in the reciprocal trade of ideas has been made, among others, by Brillouin: "We have been looking, up to now, for a physicochemical interpretation of life. It may well happen that the discovery of new laws and of some new principles in biology could result in a broad redefinition of our present laws of physics and chemistry, and produce a complete change in point of view." L. Brillouin, "Life, Thermodynamics, and Cybernetics," *American Scientist*, Vol. XXXVII (1949), p. 554. Cf. the similar prophecy of J. W. N. Sullivan, *The Limitations of Science* (New York: Mentor Books, 1948), pp. 187 ff.

15. Pascual Jordan, *Physics of the Twentieth Century* (New York: Philosophical Library, 1944), p. 151. Cf. von Bertalanffy, *Problems of Life*, pp. 165 ff.

16. *Physics of the Twentieth Century*, p. 152. Elsewhere Jordan has written: ". . . when we look more sharply, when we observe the individual atoms that make up the huge bodies and quantities of matter, we find on all sides free individual decisions which are not determined by natural law. . . . Organic life partakes then of the same freedom and spontaneity that physicists have found at the root of material being. . . . We may say then that the attempt to prove man a machine, to deny him free will, has been refuted by the sheer facts of science." *Science and the Course of History* (New Haven: Yale University Press, 1955), pp. 112–113.

17. *Physics of the Twentieth Century*, pp. 153–154. Cf. L. de Broglie, *Physics and Microphysics* (New York: Harper Torchbooks, 1960), pp. 140–141.

18. See Niels Bohr, *Atomic Physics and Human Knowledge* (New

York: Wiley, 1958), pp. 10, 62f., 96, 100f.; Werner Heisenberg, *Physics and Philosophy* (New York: Harper, 1958), esp. pp. 106 ff.; Erwin Schrödinger, *What is Life?*, (New York: Doubleday Anchor Books, 1956), pp. 2 ff. See also C. F. von Weizsäcker, *The History of Nature* (Chicago: University of Chicago Press, 1949), p. 129, *passim*.

19. Harold G. Wolff, "The Mind-Body Relationship," in Lyman Bryson (ed.), *An Outline of Man's Knowledge of the Modern World* (Garden City, N.Y.: Nelson Doubleday, 1960), p. 52.

20. *Problems of Life*, pp. 18–19. Three influential early contributors to the development of the holistic-organismic viewpoint deserve specific mention: J. S. Haldane, whose views were first formulated before the turn of the century and are best set forth in *The Philosophical Basis of Biology* (London: Hodder and Stoughton, 1931); the soldier-statesman Jan Christiaan Smuts, by virtue of his single systematic work, *Holism and Evolution* (New York: Macmillan, 1926; republished by Viking Press, Compass Books edition, 1961); and the biologist and neuropsychiatrist Kurt Goldstein, whose theories were made available in English with the translation of *The Organism: A Holistic Approach to Biology* (New York: American Book Company, 1939). For the considerable company of others, early and late, who have reinforced this theoretical approach, see von Bertalanffy, *Problems of Life*, p. 182.

21. "The intrinsic directiveness of many vital processes is as well established by factual evidence as anything in biology. To deny it simply because we do not understand it is a futile device as old at least as the Greek Eleatics and as modern as Mary Baker Eddy." C. Judson Herrick, *The Evolution of Human Nature* (Austin: University of Texas Press, 1956), p. 53. Cf. E. S. Russell, *The Directiveness of Organic Activities* (Cambridge, England: Cambridge University Press, 1945).

22. Wolff, "The Mind-Body Relationship," p. 51.

23. Edmund W. Sinnott, *Matter, Mind and Man* (New York: Harper, 1957), p. 42. Cf. McKinley, *Evolution*, pp. 44 ff.

24. Quoted in Sullivan, *The Limitations of Science*, p. 126.

25. *Ibid.*

26. Herrick, *The Evolution of Human Nature*, p. 48. Cf. Haldane, *The Philosophical Basis of Biology*, pp. 31, 43 *passim*.

27. Brillouin, "Life, Thermodynamics, and Cybernetics," p. 554.

28. Introduction to Pierre Teilhard de Chardin, *The Phenomenon of Man* (New York: Harper, 1959), pp. 26–27. Herrick has similarly written: "In the organic realm all growth and all progressive evolution manifest an apparent reversal of entropy. . . . It may well be that the reversal of entropy is true for the cosmos as a whole and that the process of degradation that we call entropy is merely a local and transient episode in a vast

domain of creative process that is continuously enlarging and progressively differentiating." *The Evolution of Human Nature*, p. 51. See also von Bertalanffy, *Problems of Life*, pp. 112 ff.; Schrödinger, *What is Life?*, pp. 69 ff.

29. For a formidable accumulation of evidence and persuasive argument supporting this point, see Michael Polanyi, *Personal Knowledge* (Chicago: University of Chicago Press, 1958), pp. 335 ff.

30. Hans Spemann, *Embryonic Development and Induction* (New Haven: Yale University Press, 1938), p. 371. (Emphasis added.)

31. *Personal Knowledge*, p. 342. Cf. Goldstein, *The Organism*, pp. 208–209.

32. *Personal Knowledge*, p. 340. Throughout his long career, J. S. Haldane gave major stress to what he termed "the axiom of life": i.e., the a priori awareness that a thing is *alive* which is given to us prior to its scientific analysis—and which such analysis can never disclose. Haldane's sustained polemics against mechanistic biology centered largely around his deep conviction of the validity and viability for the life sciences of this "subjective" factor in observation. See his *The Philosophical Basis of Biology*, pp. 14 ff., 18 ff. Cf. Haldane, *The Philosophy of a Biologist* (Oxford: Clarendon Press, 1936), pp. 73 ff.; *Materialism* (London: Hodder and Stoughton, 1932), pp. 64 ff.

33. The recognition of an organism "in itself," or as a whole, in contradistinction to the particulate analysis of organic processes, has been conveyed somewhat differently by von Bertalanffy: "There is a kind of complementarity between analytical and global treatment of biological systems. We can either pick out individual processes in the organism and analyse them in terms of physico-chemistry—then the whole, because of its enormous complexity, will escape us; or we can state global views for the biological system as a whole—but then we have to forsake the physico-chemical determination of the individual processes." *Problems of Life*, p. 155.

34. Ralph S. Lillie, *General Biology and Philosophy of Organism* (Chicago: University of Chicago Press, 1945), p. 50. This entire book may be regarded as an elaboration and documentation of the physical-psychical expression of complementarity. Cf. Lillie, "Biological Causation," *Philosophy of Science*, Vol. VII (1940), pp. 315–316, 326, 327.

35. *Problems of Life*, p. 173.

36. *Ibid.*, p. 175. (Emphasis added.)

37. Augusto Pi Suner, *The Bridge of Life* (New York: Macmillan, 1951), pp. 171–172. On the concept of spontaneity, see also Paul Weiss, *Nature and Man* (New York: Henry Holt, 1947), pp. 58 ff.

38. "In biology it may be important for a complete understanding

that the questions are asked by the species man which itself belongs to the genus of living organisms, in other words, that we already know what life is even before we have defined it scientifically." Heisenberg, *Physics and Philosophy*, p. 107.

39. Thus N. J. Berrill observes that "the greatest scientific insights, which have survived the most rigorous scrutiny, have come when the mind has virtually been one with what it contemplates, when self as a person or a human being is submerged and a feeling of identity with the subject takes its place. . . . If you bring your mind and senses close enough and long enough to any creature from jellyfish to monkey, sooner or later it begins to explain itself in a manner you can understand but may have trouble in translating into language. In some degree I know this to be true." *Man's Emerging Mind* (New York: Dodd, Mead, 1955), p. 280.

40. *The Phenomenon of Man*, p. 56.

41. *Ibid.*, p. 53. The comment of Julian Huxley, in his introduction to this remarkable book, is pertinent here: "This view admittedly involves speculation of great intellectual boldness, but the speculation is extrapolated from a massive array of fact, and is disciplined by logic. It is, if you like, visionary; but it is the product of a comprehensive and coherent vision." *Ibid.*, p. 16.

42. Loren Eiseley, *The Firmament of Time* (New York: Atheneum Publishers, 1960), p. 141.

43. See, for example, Julian Huxley's introduction to the Mentor edition of *The Origin of Species* (New York: 1958), p. xv.

44. Cf. George Gaylord Simpson, *The Meaning of Evolution* (New York: Mentor Books, 1951), p. 147; McKinley, *Evolution*, pp. 223–225.

45. On the role of cooperation, and its biological basis, see especially M. F. Ashley Montagu, *The Direction of Human Development* (New York: Harper, 1955), pp. 17–59; and the same author's *Darwin, Competition, and Cooperation* (New York: Abelard-Schuman, 1952).

46. V. G. Childe, *Man Makes Himself* (New York; Mentor Books, 1951).

47. Simpson, *The Meaning of Evolution*, p. 140.

48. Cf. Julian Huxley, *Evolution in Action* (New York: Mentor Books, 1953), pp. 125 ff.; E. W. Sinnott, "The Biological Basis of Democracy," *Yale Review*, Vol. XXXV (1945), pp. 61–73; C. H. Waddington, "Human Ideals and Human Progress," *World Review* (August, 1946), pp. 29–36; T. Dobzhansky, *The Biological Basis of Human Freedom* (New York: Columbia University Press, 1956), pp. 80 ff., 111 ff.; C. D. Leake, "Ethicogenesis," *Scientific Monthly*, Vol. LX (1945), pp. 245–253; P. B. Medawar, *The Uniqueness of the Individual* (New York: Basic Books, 1957), pp. 138 ff.

49. *The Meaning of Evolution*, pp. 141–142.

50. McKinley, *Evolution*, p. 69.

51. As the new evolution reveals itself principally through history, it is fitting that it should have been a distinguished student of history, Croce, who has emphasized this point in his *History as the Story of Liberty* (New York: Norton, 1941). His title and theme, Croce explains, are drawn from the famous statement of Hegel, the meaning of which is simply that "liberty is the eternal creator of history and itself the subject of every history. As such it is on the one hand the explanatory principle of the course of history, and on the other the moral ideal of humanity." *Ibid.*, p. 59.

52. "Biology thus reinstates man in a position analogous to that conferred on him as Lord of Creation by theology." Julian Huxley, *Man in the Modern World* (New York: Mentor Books, 1948), p. 9.

53. "In broad terms, the destiny of man on earth has been made clear by evolutionary biology. It is to be the agent of the world process of evolution, the sole agent capable of leading it to new heights, and enabling it to realize new possibilities." Julian Huxley, *Evolution in Action*, p. 31. Cf. Huxley, *Man in the Modern World*, p. 27; McKinley, *Evolution*, p. 239; Dobzhansky, *The Biological Basis of Human Freedom*, p. 88.

54. *The Meaning of Evolution*, p. 181.

55. See the argument of Paul Weiss for freedom as the fundamental explanatory factor underlying organic evolution. E.g.: "Evolution does not apply merely to parts of beings or to their bodies. It embraces the living being as a whole—its sensitivity, its concern and its end. It is a product of freedom. Freedom is the power behind evolution, responsible for whatever mutations occur and for the fact that higher beings have non-bodily powers such as sensitivity and purposiveness." *Nature and Man*, p. 102, *passim*.

56. Herrick, *The Evolution of Human Nature*, pp. 288 ff.

CHAPTER VI   *The Freedom to Be Human*

1. C. Judson Herrick, *George Ellet Coghill: Naturalist and Philosopher* (Chicago: University of Chicago Press, 1949), p. 137, *passim*. Cf. Coghill, *Anatomy and the Problem of Behavior* (London: Cambridge University Press, 1929).

2. C. Judson Herrick, *The Evolution of Human Nature* (Austin: University of Texas Press, 1956), pp. 452–453.

3. *Ibid.*, pp. 292, 296.

4. See Alfred Lief, *The Commonsense Psychiatry of Dr. Adolf Meyer* (New York: McGraw-Hill, 1948), p. 387.

5. Herrick, *The Evolution of Human Nature*, p. 303.

6. "Feelings may be called 'subjective' because they belong to the subject or because they are exposed to the subjectivity of the observer. In the first sense they belong to the object under observation, which in this case is man as subject. The second meaning cannot impair the first: science must recognize feelings as part of the objective reality though it may be difficult to eliminate the subjectivity of the observer. Science begins by respecting the subject matter." Kurt Riezler, *Man: Mutable and Immutable* (Chicago: Regnery, 1950), p. 112.

7. Herrick, *The Evolution of Human Nature*, p. 304.

8. Sir Charles Sherrington, *Man on His Nature* (New York: Doubleday Anchor Books, 1953), pp. 297–298.

9. Sherrington, *The Brain and Its Mechanism* (London: 1936).

10. *Man on His Nature*, p. 252.

11. *Ibid.*, p. 251.

12. "An entity which if it were accessible to experience is not open to quantitative measure, and otherwise is not open to experience even at all, is an entity indeed refractory to analytic treatment. . . . Conscious experience seems refractory to measurement in terms of itself." *Ibid.*, pp. 250–251.

13. *Ibid.*, p. 255.

14. *Ibid.*, p. 258. Cf. Pierre Teilhard de Chardin, *The Phenomenon of Man* (New York: Harper, 1959), Book Three: "Thought."

15. Sherrington, *Man on His Nature*, p. 260.

16. *Ibid.*, p. 261.

17. *Ibid.*

18. This suggestion of the harmonization at a higher level of the two "natures" of man (physical and mental, objective and subjective) parallels the illuminating treatment of "double-boundary concepts" which runs through the philosophical writings of William Ernest Hocking. Such a concept, he points out, is that of "Nature"—which in its narrower sense is clearly set apart from the human mind but "in its wider sense includes its own narrower sense together with the opposite of the narrower"—i.e., mental experience. Hocking, "Response to Professor Krikorian's Discussion," *Journal of Philosophy*, Vol. LV, No. 7 (March 27, 1958), pp. 278–279.

19. Cf. George S. Klein, "Perception, Motives and Personality," in J. L. McCary (ed.), *Psychology of Personality* (New York: Grove Press, 1956), pp. 123 ff.

20. For a brief and trenchant criticism of this traditional approach, see H. A. Witkin (*et al.*), *Personality Through Perception* (New York: Harper, 1954), pp. 2 f., *passim.*

21. W. H. Ittelson, *Visual Space Perception* (New York: Springer Publishing Co., 1960), p. 18. Cf. W. H. Ittelson and Hadley Cantril, *Perception: A Transactional Approach* (Garden City, N. Y.: Doubleday, 1954), p. 3.

22. "As one knows, Gestalt psychology has made tremendous contributions in this area. However, for the most part, Gestalt psychology handled the problem of configuration in terms of structure rather than dynamics. The problems of Gestalt psychology were mainly those of part and whole, reorganization, and formal laws of organization such as autochthonous laws. Wherever the problem of perceptual dynamics was treated, it was either in terms of phenomeno-logical analysis or in terms of electro-physiological or electro-chemical models." Heinz Werner and Seymour Wapner, "Toward a General Theory of Perception," in David C. Beardslee and Michael Wertheimer, *Readings in Perception* (New York: Van Nostrand, 1958). Cf. the famous critique of the Gestalt theory of perception ("no Gestalt without a Gestalter") by William Stern, *General Psychology from the Personalistic Standpoint* (New York: Macmillan, 1933), pp. 114 ff.

23. D. W. Hamlyn, in an incisive study bearing the subtitle "A Philosophical Examination of Gestalt Theory and Derivative Theories of Perception," notes that Gestaltists have given some attention to individual differences and subjective factors such as past experience, as in Werthei- mer's Laws of Gestalt Form. "But such effects have been played down. . . . The general theory does not allow for these effects. For what the theory says is that *whatever* we see is a product of the interaction between external forces set up by stimuli from the object and autonomous internal forces in the cortex, so that we see things as wholes and as wholes with the simplest or 'best' structure. The apparatus of the theory is so far purely physiological and perception is regarded merely as the end-product of such processes of neural excitation." *The Psychology of Perception* (London: Routledge and Kegan Paul, 1957), p. 58.

24. *Ibid.*, p. 63. Cf. the similar statement by Gordon W. Allport, *Personality and Social Encounter* (Boston: Beacon Press, 1960), pp. 25–26.

25. A fairly representative recent statement is that of Franklin P. Kilpatrick: "This means . . . that perception cannot be 'due to' the physiological stimulus pattern; . . . There must be, in addition, some basis for the organism's 'choosing' one from among the infinity of external conditions to which the pattern might be related. Thus, any notion concerning a unique correspondence between percept and object must be abandoned, and a discovery of the factors involved in the 'choosing' activity of the organism becomes the key problem in perceptual theory." Kilpatrick

(ed.), *Explorations in Transactional Psychology* (New York: New York University Press, 1961), p. 3.

26. See Harry Stack Sullivan, *The Interpersonal Theory of Psychiatry* (New York: Norton, 1953), pp. 170, 233 f. Cf. Mason Haire and Willa Freeman Grunes, "Perceptual Defenses: Processes Protecting an Organized Perception of Another Personality," *Human Relations*, Vol. III, No. 4 (December, 1950), pp. 403–412. See also Franklin Fearing, "Toward a Psychological Theory of Human Communication," *Journal of Personality*, Vol. XXII, No. 1 (September, 1953), pp. 71–88.

27. Muzafer Sherif, *An Outline of Social Psychology* (New York: Harper, 1948), pp. 217 f., 221 f.

28. L. Postman, in J. H. Rohrer and M. Sherif, *Social Psychology at the Crossroads* (New York: Harper, 1951), chap. 10. Cf. Floyd H. Allport, *Theories of Perception and the Concept of Structure* (New York: Wiley, 1955), pp. 375 ff.

29. *Ibid.*

30. Cf. R. R. Blake and J. M. Vanderplas, "The Effect of Prerecognition Hypotheses on Veridical Recognition Thresholds in Auditory Perception," *Journal of Personality*, Vol. XIX (1950), pp. 95–115; G. S. Klein, H. J. Schlesinger and D. E. Meister, "The Effect of Personal Values on Perception," *Psychological Review*, Vol. LVIII (1951), pp. 96–112; D. McClelland and J. Atkinson, "The Projective Expression of Needs," *Journal of Psychology*, Vol. XXV (1948), pp. 205–222. Cf. Benjamin G. Rosenberg, *Compulsiveness as a Determinant in Selected Cognitive-Perceptual Performances* (unpublished dissertation, University of California, 1952), p. 60.

31. Cf. Else Frenkel-Brunswik, "Tolerance Toward Ambiguity as a Personality Variable" (Abstract), *American Psychologist*, Vol. III (1948), p. 268.

32. Cf. Gordon W. Allport, *Becoming* (New Haven: Yale University Press, 1955), p. 9, *passim.*

33. The best historical discussion is still that of E. G. Boring, *A History of Experimental Psychology* (New York: Century Co., 1929), pp. 539 ff. Cf. the interesting treatment of functionalism by Richard Muller-Freienfels, *The Evolution of Modern Psychology* (New Haven: Yale University Press, 1935), pp. 238 ff. See also R. S. Woodworth, *Contemporary Schools of Psychology* (New York: Ronald Press, 1948), chap. 2.

34. The specific source is John Dewey and Arthur F. Bentley, *Knowing and the Known* (Boston: Beacon Press, 1949). Dewey's earlier "interactionist" conception, of course, anticipated the later formulation. See Ittelson and Cantril, *Perception*, p. 3.

35. Dewey, "The Reflex Arc Concept in Psychology," *Psychological Review*, Vol. III (1896), pp. 357–370.

36. Dewey, *Logic: The Theory of Inquiry* (New York: Henry Holt, 1938), p. 30.

37. G. H. Mead, *Philosophy of the Act* (Chicago: University of Chicago Press, 1938), p. 654. Elsewhere Mead wrote: "The intelligence that is involved in perception is elaborated enormously in what we call 'thought.' One perceives an object in terms of his response to it. . . . It is true of all our experience that it is the response that interprets to us what comes to us in the stimulus, and it is such attention which makes the percept out of what we call 'sensation.' The interpretation of the response is what gives the content to it." *Mind, Self and Society* (Chicago: University of Chicago Press, 1931), p. 114.

38. Ittelson, *Visual Space Perception*, p. 23. For similar statements, cf. Lawrence E. Cole, *Human Behavior: Psychology as a Bio-Social Science* (1953), p. 358; Lawrence K. Frank, *Individual Development* (New York: Doubleday, 1955), pp. 23 ff.

39. See Hadley Cantril and Charles H. Bumstead, *Reflections on the Human Venture* (New York: New York University Press, 1960), p. 49.

40. Ittelson, *Visual Space Perception*, p. 23. Cf. the statement of an existential psychologist, Werner Wolff: "The perception of reality is therefore an evaluation of reality. . . . Man chooses his space and he accentuates his time. Like an artist, man selects his material of reality. This selection of reality material is the first step in the creative act of perception." *Values and Personality: An Existential Psychology of Crisis* (New York: Grune & Stratton, 1950), p. 19.

41. Ittelson and Cantril, *Perception*, p. 7. Cf. Cantril and Bumstead, *Reflections on the Human Venture*, p. 59: "Attention is always a function of intention. . . . While past and present are involved in perceiving, its chief value comes from its orientation toward the future."

42. Charles M. Solley and Gardner Murphy, *Development of the Perceptual World* (New York: Basic Books, 1960), p. 290.

43. *Ibid.*

44. *Ibid.*

45. Gordon W. Allport, "The Trend in Motivational Theory," *American Journal of Orthopsychiatry* (1953), reprinted in Allport, *Personality and Social Encounter* (Boston: Beacon Press, 1960), p. 95.

46. *Ibid.*, p. 97.

47. *Ibid.*, p. 99.

48. *Ibid.*, p. 98.

49. William H. Whyte, Jr., *The Organization Man* (New York: Simon and Schuster, 1956), p. 40.

50. The point has been lucidly made by W. Macneile Dixon: "En-

quire of the philosophers, and they will tell you that this matter of perception is still more, indeed desperately complicated. They may even assure you that there is no such thing as pure perception, that it has no story of its own to tell, that it is a part, and a part only, of that active and energetic faculty we call thought, and that it involves memory and anticipation, forethought and afterthought. . . . In a word, to separate the seen from the seer is allowed to be finally and utterly impossible. We are, as Niels Bohr has expressed it, 'both spectators and actors in the great drama of existence.' . . . The senses, then, do not give us knowledge. An active, inner principle, wholly independent of the senses is essential to the process of obtaining it." *The Human Situation* (New York: St. Martin's Press, n.d.), pp. 58–59.

51. Cantril and Bumstead, *Reflections on the Human Venture*, p. 43.

52. Cf. Ittelson, *Visual Space Perception*, p. 21.

53. "I shall designate these two basic modes of perceptual relatedness as the subject-centered, or *autocentric*, and the object-centered, or *allocentric*, mode of perception. The main differences between the autocentric and allocentric modes of perception are these: In the autocentric mode there is little or no objectification; the emphasis is on how and what the person feels; there is a close relation, amounting to a fusion, between sensory quality and pleasure or unpleasure feelings, and the perceiver reacts primarily to something impinging on him. . . . In the allocentric mode there is objectification; the emphasis is on what the object is like . . . the perceiver usually approaches or turns to the object actively and in doing so either opens himself toward it receptively or, figuratively or literally, takes hold of it, tries to 'grasp' it." Ernest G. Schachtel, *Metamorphosis* (New York: Basic Books, 1959), p. 83.

54. Cf. Solley and Murphy, *Development of the Perceptual World*, p. 292.

55. *Ibid.*, pp. 292–293. (Emphasis added.) Similarly, Werner Wolff describes "two sets of perception which differ from each other because of the two sets of experience the perceptions are based upon." The first he illustrates in the experience of perceiving road *signs;* the second in the perception of *scenery.* "In the road-sign-perception-and-experience, man reacts like a static response mechanism; his reaction is more or less stereotyped and can be predicted. The response mirrors the stimulus. In the scenery-perception-and-experience, man reacts as a dynamic organism; his reaction is individualized and cannot be predicted. His response transforms the stimulus, and this transformation is a creative act." *Values and Personality*, pp. 31, 32.

56. Solley and Murphy, *Development of the Perceptual World*, pp. 303–304. Cf. Ittelson and Cantril, *Perception*, p. 27: "The primary function of perception is neither revelation of the present nor remembrance of the past: it is prediction of the future."

impulses were visualized as 'a network of communicating canals filled with liquid,' and the libido was *something* '(an amount of affect, a sum of excitation), something having all the attributes of a quantity—although we possess no means of measuring it—a something which is capable of increase, decrease, displacement and discharge, and which extends itself over the memory-traces of an idea like an electric charge over the surface of the body.' " Floyd W. Matson, "The Political Implications of Psychoanalytic Theory," *Journal of Politics*, Vol. XVI, No. 4 (November, 1954), pp. 705–706. See Freud, *Collected Papers* (London: Hogarth Press, 1924), Vol. I, p. 75. Cf. Clara Thompson, *Psychoanalysis: Evolution and Development* (New York: Grove Press, 1957), pp. 18 ff.; Patrick Mullahy, *Oedipus: Myth and Complex: A Review of Psychoanalytic Theory* (New York: Hermitage Press, 1948), pp. 318 ff.

70. Thus Jones points out that Freud was "ill-informed in the field of contemporary psychology and seems to have derived only from hearsay any knowledge he had of it. He often admitted his ignorance of it, and even when he tried to remedy it later did not find anything very useful for his purpose in it, with perhaps two exceptions. . . . Not having been schooled in any psychological discipline, even in the little there was at that period, he was apt to be careless and imprecise in his use of terms. . . ." *Life and Work of Sigmund Freud*, Vol. I, p. 371.

71. Sigmund Freud, *New Introductory Lectures on Psychoanalysis* (New York: Norton, 1933), p. 103.

72. Karl Stern, *The Third Revolution* (New York: Harcourt, Brace, 1954), pp. 154 ff.

73. See Jones, *Life and Work of Sigmund Freud*, Vol. I, pp. 366 ff.

74. Compare the statement of Rank: ". . . the ego, instead of being caught between the two powerful forces of fate, the inner id and the externally derived super-ego, develops and expresses itself creatively. The Freudian ego driven by the libidinal id and restrained by parental morality, becomes almost a nonentity, a helpless tool for which there remains no autonomous function. . . . In my view the ego is much more than a mere show place for the standing conflict between two great forces." Otto Rank, *Will Therapy and Truth and Reality* (New York: Knopf, 1945), p. 212.

75. Sigmund Freud, *The Future of an Illusion* (New York: Doubleday Anchor Books, n.d.), p. 4.

76. Sigmund Freud, *Civilization and Its Discontents* (London: Hogarth Press, 1930), p. 105. Elsewhere Freud wrote that "throughout the life of the individual there is a constant replacement of the external compulsion by the internal." "Thoughts for the Times on War and Death," in *Civilization, War, and Death*, p. 9.

77. *Civilization and Its Discontents*, pp. 102–103.

78. David Bakan, *Sigmund Freud and the Jewish Mystical Tradi-*

57. Ittelson and Cantril, *Perception*, p. 7. The same authors state: "It is not necessary, at this point, to argue the question of determinism or indeterminism in the 'real' world, but simply to point out that the world of experience is, to a greater or lesser extent, always indeterminate. Every immediate concrete situation we face has a certain degree of novelty. . . . To this extent it is unpredictable, indeterminate . . ." *Ibid.*, p. 29.

58. Harold Grier McCurdy, *The Personal World: An Introduction to the Study of Personality* (New York: Harcourt, Brace & World, 1961), p. 581.

59. C. F. von Weizsäcker, *The World View of Physics* (Chicago: University of Chicago Press, 1952), p. 206. The full statement is: "But the fact that science does not know the 'Thou' leads us to an entirely different question: not whether one *can* experiment with man, but whether one wants to or may experiment with him. What do I do to my fellow man in treating him in my thoughts or actions as a mere object? In this question seems to focus everything that has rightly been advanced against the application of natural science to man."

60. Schachtel, *Metamorphosis*, p. 238. A. H. Maslow has dealt with this phenomenon under the label of "rubricizing perception," which he describes as "a classifying, ticketing, or labeling of the experience rather than an examination of it . . . What we do in stereotyped or rubricized perceiving is parallel to the use of clichés and hackneyed phrases in speaking." *Motivation and Personality* (New York: Harper, 1954), p. 268.

61. Schachtel, *Metamorphosis*, p. 248.

62. This conception, central to existentialist philosophy from Kierkegaard to Sartre, finds an equivalent expression in present-day psychologies of self-affirmation and conscious choosing—some of which are discussed below.

63. Ittelson and Cantril, *Perception*, p. 31.

64. *The Origins of Psychoanalysis: Sigmund Freud's Letters* (New York: Basic Books, 1954), p. 355. See above, Chapter I.

65. See Ernest Jones, *Life and Work of Sigmund Freud* (New York: Basic Books, 1953), Vol. I, esp. Chap. XVII.

66. *Ibid.*, Vol. I, p. 34.

67. *Ibid.* (Emphasis added.)

68. *Ibid.*, Vol. I, p. 40.

69. Many of these concepts, to quote from an earlier essay of the present writer, were "hypostatized as substantial, quantitative forces, and rendered in a vivid mechanical imagery of push-pull, ebb and flow, dam and release—much of which was meant quite literally. To Freud an instinct was a quantum of energy forcing its way in one direction or another; . . . sexual

tion (New York: Van Nostrand, 1958), p. 219. "As we hope we can demonstrate, in the same way that the image of Moses was for Freud both a reality and a fiction, so was the image of the Devil." *Ibid.*, p. 188.

79. Norman O. Brown, *Life Against Death* (Middletown, Conn.: Wesleyan University Press, 1959), p. 10. Cf. Herbert Marcuse, *Eros and Civilization* (Boston: Beacon Press, 1951). Thus Marcuse has written: "The social content of Freudian theory becomes manifest: sharpening of the psychoanalytical concepts means sharpening their critical function, their opposition to the prevailing form of society. And this critical sociological function of psychoanalysis derives from the fundamental role of sexuality as a 'productive force', the libidinal claims propel progress toward freedom and universal gratification of human needs beyond the patricentric-acquisitive stage." Marcuse, "The Social Implications of Freudian Revisionism," *Dissent* (Summer, 1955), reprinted in *Voices of Dissent* (New York: Grove Press Evergreen Books, 1958), p. 296.

80. See Erich Fromm, *Escape from Freedom* (New York: Farrar & Rinehart, 1941), pp. 10–11. Cf. Clara Thompson, *Psychoanalysis: Evolution and Development*, chap. 7, "Freud's Cultural Orientation Compared with Modern Ideas of Culture."

81. Cf. Richard T. LaPiere, *The Freudian Ethic* (New York: Duell, Sloane and Pearce, 1959).

82. Robert M. Lindner, *Prescription for Rebellion*.

83. Jacques Maritain, "Freudianism and Psychoanalysis: A Thomist View," in Benjamin Nelson (ed.), *Freud and the Twentieth Century* (New York: Meridian Books, 1957), pp. 248–249.

84. In addition to the well-known expression of this mood in *Civilization and Its Discontents*, see Freud's exchange of letters with Albert Einstein, *Why War?* reprinted in *Collected Papers of Sigmund Freud* (New York: International Psychoanalytic Press), Vol. V (1950), pp. 273–287.

85. Reinhold Niebuhr, *The Nature and Destiny of Man* (New York: Scribner's, 1948), Vol. I, p. 42. Cf. Lionel Trilling, *The Liberal Imagination* (New York: Viking, 1950), pp. 35, 36.

86. *The Thomas Mann Reader* (New York: Knopf, 1950), p. 454.

87. See, for example, Stanley Edgar Hyman, "Psychoanalysis and the Climate of Tragedy," in *Freud and the Twentieth Century*, pp. 167–185.

88. Cf. Matson, "The Political Implications of Psychoanalytic Theory," pp. 722–723; Will Herberg, "Freud, the Revisionists, and Social Reality," in *Freud and the Twentieth Century*, p. 153.

89. Helen Merrell Lynd, "Must Psychology Aid Reaction?" *The Nation*, Vol. CLXVIII, No. 3 (January 15, 1949), pp. 76 ff.

90. Cf. Daniel Bell, *The End of Ideology: On the Exhaustion of Political Ideas in the Fifties* (Glencoe, Ill.: Free Press, 1960). Bell notes that "the basic political drift of the former Left intelligentsia in the United States in the forties and fifties has been anti-ideological—that is to say, skeptical of the rationalistic claim that socialism, by eliminating the economic basis of exploitation, would solve all social questions; and to a great extent this anti-rationalism is the source of the intellectual vogue of Freudianism and neo-orthodox theology (i.e., Reinhold Niebuhr and Paul Tillich)." *Ibid.*, p. 297.

91. *Life Against Death*, pp. ix–x.

92. *Ibid.*, p. x.

93. Thus Marcuse in particular justifies his avoidance of therapeutic considerations on the basis of "a discrepancy between theory and therapy inherent in psychoanalysis itself"—which is that "while psychoanalytic theory recognizes that the sickness of the individual is ultimately caused and sustained by the sickness of his civilization, psychoanalytic therapy aims at curing the individual so that he can continue to function as part of this civilization without surrendering to it altogether." *Voices of Dissent*, p. 207. Cf. Brown, *Life Against Death*, p. xi.

94. Maritain, "Freudianism and Psychoanalysis," p. 230.

95. The phrase is that of Patrick Mullahy; but it is echoed in nearly identical terms throughout the unorthodox literature. Mullahy, *Oedipus: Myth and Complex*, p. 320.

96. Cf. Fay B. Karpf, *The Psychology and Psychotherapy of Otto Rank* (New York: Philosophical Library, 1953), p. 13, *passim*.

97. On the original debates between Adler and Freud, see the contribution of Kenneth Mark Colby in Heinz and Rowena Ansbacher, *The Individual Psychology of Alfred Adler* (New York: Basic Books, 1956), pp. 69 ff.

98. An illuminating discussion of Adler's holistic orientation, in relation to biological theory, is to be found in Lewis Way, *Adler's Place in Psychology* (London: Allen & Unwin, 1950), chap. 2, "Wholeness and Purpose."

99. Clara Thompson, *Psychoanalysis: Evolution and Development*, p. 11.

100. Thus Dreikurs writes: "The holistic view of man is reflected in the name, Individual Psychology. It was meant to indicate the indivisibility of man, . . . In contrast to prevalent mechanistic-deterministic theories, Adler recognized man's creative ability, which permits him to set his own goals and determine his own movements. His goals express his total personality, his past, his present and his move toward the future." Rudolf Dreikurs, "Adlerian Psychotherapy," in F. Fromm-Reichmann and J. L. Morene

(eds.), *Progress in Psychotherapy: 1956* (New York: Grune & Stratton, 1956), p. 111.

101. Alfred Adler, *The Practice and Theory of Individual Psychology* (New York: Harcourt, Brace, 1927), p. 244.

102. Ansbacher, *The Individual Psychology of Alfred Adler*, p. 334.

103. *Ibid.*, pp. 340–341.

104. For accounts of various of these connections and derivations, see the following articles in Kurt A. Adler and Danica Deutsch (eds.), *Essays in Individual Psychology* (New York: Grove Press, 1959): Lucia Radl, "Existentialism and Adlerian Psychology"; Robert W. White, "Adler and the Future of Ego Psychology"; H. L. Ansbacher, "Causality and Indeterminism, According to Alfred Adler, and Some Current American Personality Theories."

105. Thompson, *Psychoanalysis: Evolution and Development*, p. 169.

106. Jolan Jacobi, *The Psychology of Jung* (New Haven: Yale University Press, 1943), p. 88. Cf., on this point, Calvin S. Hall and Gardner Lindzey, *Theories of Personality* (New York: Wiley, 1957), p. 78.

107. C. G. Jung, *Modern Man in Search of a Soul* (New York: Harcourt, Brace Harvest Books, 1960), p. 120.

108. *Ibid.*, p. 121.

109. *Ibid.*, p. 26.

110. C. G. Jung, *The Undiscovered Self* (New York: Mentor Books, 1959), p. 18. "Under the influence of scientific assumptions, not only the psyche but the individual man and, indeed, all individual events whatsoever suffer a leveling down and a process of blurring that distorts the picture of reality into a conceptual average. We ought not to underestimate the psychological effect of the statistical world picture: it displaces the individual in favor of anonymous units that pile up into mass formations." *Ibid.*, p. 21.

111. "If I want to understand an individual human being I must lay aside all scientific knowledge of the average man and discard all theories in order to adopt a completely new and unprejudiced attitude. I can only approach the task of *understanding* with a free and open mind . . ." *Ibid.*, p. 18.

112. *Ibid.*, pp. 18–19.

113. *Modern Man in Search of a Soul*, p. 49.

114. *Ibid.*, p. 50. Cf. C. G. Jung, *Essays on Contemporary Events* (London: Kegan Paul, 1947), pp. 39 f.

115. *Modern Man in Search of a Soul*, p. 53.

116. "The uniqueness of the individual and of his situation stares the doctor in the face and demands an answer. . . . As understanding deepens, the further removed it becomes from knowledge." *The Undiscovered Self*, pp. 62–63.

117. *Modern Man in Search of a Soul*, p. 53.

118. *Modern Man in Search of a Soul*, p. 54.

119. *Ibid.*, p. 41.

120. On Jung's "law of opposites," see *The Basic Writings of C. G. Jung*, edited by V. S. de Laszlo (New York: Modern Library, 1959), p. 347, *passim*.

121. Cf. Ira Progoff, *Jung's Psychology and Its Social Meaning* (New York: Grove Press, 1955), pp. 59 f.

122. *Modern Man in Search of a Soul*, p. 122.

123. Jacobi, *The Psychology of Jung*, p. 68. Cf. Jung's discussion of complementarity in physics and psychology, *The Basic Writings of C. G. Jung*, pp. 102 f.

124. *Jung's Psychology and Its Social Meaning*, p. 290.

125. See Karpf, *The Psychology and Psychotherapy of Otto Rank*, chap. 1.

126. See Otto Rank, *Will Therapy and Truth and Reality* (New York: Knopf, 1945), p. 5.

127. *Ibid.*, p. 2.

128. Otto Rank, *Beyond Psychology* (New York: Dover, 1958), p. 290. Elsewhere he wrote: "Here psychotherapy enters as a binding function, not only in its effort to bind the isolated neurotic to society, but even in its method which offers to the patient in the person of the analyst, the 'thou' from whom he had estranged himself in self-willed independence." *Will Therapy and Truth and Reality*, p. 155.

129. For example, he told an international congress of mental hygiene workers in 1930: "This conflict will not be lessened until we admit that science has proved to be a complete failure in the field of psychology, i.e., in the betterment of human nature and in the achievement of human happiness toward which all mental hygiene is ultimately striving." Quoted in Jessie Taft, *Otto Rank* (New York: Julian Press, 1958), p. 147.

130. *Ibid.*

131. Otto Rank, *Psychology and the Soul* (Philadelphia: University of Pennsylvania Press, 1950), p. 172.

132. *Beyond Psychology*, p. 28.

133. Compare LaPiere, *The Freudian Ethic*, cited above in connection with left anti-Freudianism.

134. *Beyond Psychology*, p. 28.

135. *Will Therapy and Truth and Reality*, pp. 219–220. See also Gerald S. Blum, *Psychoanalytic Theories of Personality* (New York: McGraw-Hill, 1953), pp. 3 ff.

136. There is a remarkable similarity between Rank's three character types and the "three types of personality" adumbrated in the 1920's by Thomas and Znaniecki: i.e., "the Philistine, the Bohemian and the creative man." W. I. Thomas and Florian Znaniecki, *The Polish Peasant in Europe and America* (New York: Dover Publications, 1958 edition), Vol. II, pp. 1831, 1837–1838, 1850 ff.

137. Cf. Karpf, *The Psychology and Psychotherapy of Otto Rank*, p. 110n.

138. Ruth Munroe, *Schools of Psychoanalytic Thought* (New York: Dryden Press, 1955), p. 584. She goes on to point out that for Rank the "personal will cannot become truly constructive until it is accepted by another person, human or divine. . . . The act of separation is not enough, no matter how heroically accomplished. . . . Rank's choice of the term *artist* for the human ideal is an attempt to convey a sense of creative integration as the highest goal of man—in contrast to more limited ideals of spiritual or material achievement." *Ibid.*, p. 586.

139. *Beyond Psychology*, pp. 47–48.

140. This title was first proposed by Martin Buber, at least indirectly. His term was "psycho-synthetic" as opposed to "psychoanalytic." See Maurice S. Friedman, *Martin Buber: The Life of Dialogue* (New York: Harper Torchbooks, 1960), p. 190.

141. Izette de Forrest, *The Leaven of Love: A Development of the Psychoanalytic Theory and Technique of Sandor Ferenczi* (New York: Harper, 1954).

142. Cf. Wilhelm Stekel, *The Interpretation of Dreams* (New York: Liveright, 1943), Vol. I. Cf. also Samuel Lowy and Emil A. Gutheil, "Active Analytic Psychotherapy (Stekel)," in *Progress in Psychotherapy: 1956*, pp. 136–143.

143. De Forrest, *The Leaven of Love*, p. 65.

144. See the account of Freud's rupture with Ferenczi in Erich Fromm, *Sigmund Freud's Mission* (New York, Harper, 1959).

145. On Sullivan, see below, pp. 232 ff.

146. This article was the presidential address at the 1943 meeting of the Eastern Psychological Association, was published in the *Psychological Review*, and is reprinted in Gordon W. Allport, *Personality and Social Encounter* (Boston: Beacon Press, 1960), pp. 71–94.

147. *Ibid.*, p. 75.

148. Robert W. White, "Adler and the Future of Ego Psychology," in *Essays in Individual Psychology*, p. 442.

149. Rank, *Will Therapy and Truth and Reality*, p. 212.

150. Cf. Munroe, *Psychoanalytic Thought*, pp. 89 ff.

151. Reprinted in David Rapaport, *Organization and Pathology of Thought* (New York: Columbia University Press, 1951), p. 383 *et seq.*

152. Rapaport summarizes Hartmann's view: "The functions which are at any given time outside the range of conflict he conceptualizes as belonging to the conflict-free ego sphere. From the role of this primary autonomy in ego development, he concluded that our usual conception that the id pre-exists the ego is inadequate, that there must be a period in human individual development in which what will later be ego, and what will later be id, co-exist . . . and that it is by the differentiation from an initial undifferentiated phase that both the ego and id arise." "The Autonomy of the Ego," in R. P. Knight and C. R. Friedman, *Psychoanalytic Psychiatry and Psychology* (New York: International Universities Press, 1954).

153. Another line of development in ego psychology paralleling that of Hartmann, Kris and others of their school is to be seen in the work of W. R. D. Fairbairn, who developed an ego-object or object-relations theory which added an accent on interpersonal relationships. See W. R. D. Fairbairn, *Psychoanalytical Studies of the Personality* (London: Tavistock, 1952). Cf. Hans Guntrip, *Personality Structure and Human Interaction* (New York: International Universities Press, 1961).

154. *Personality and Social Encounter*, p. 29.

155. *Ibid.*, p. 28.

156. See Gordon W. Allport, *Becoming, passim.*

157. Allport, *Personality and Social Encounter*, pp. 71–72.

158. Criticism to this effect is to be found in Patrick Mullahy, *Oedipus: Myth and Complex*, pp. 333 f.

159. Erich Fromm, *The Art of Loving* (New York: Harper & Row, 1957).

160. Karen Horney, *Neurosis and Human Growth* (New York: Norton, 1950), p. 368.

161. Cf. Alexander Reid Martin, "The Whole Patient in Therapy," in *Progress in Psychotherapy: 1956*, p. 174.

162. See Karen Horney, *New Ways in Psychoanalysis* (New York: Norton, 1939), pp. 8, 9, 285, *passim.*

163. See, in particular, Erich Fromm, *Man for Himself* (New York: Rinehart, 1947), chap. 3, 4.

164. Norman O. Brown, *Life Against Death*, p. x.

165. Cf. Horney, *The Neurotic Personality of Our Time* (New York: Norton, 1937); Fromm, *Escape from Freedom, passim.*

166. Cf. Mullahy, *Oedipus: Myth and Complex*, sections on Sullivan in relation to the neo-Freudians.

167. This was Freud's basic formulation of the theme; but for its extrapolation into culture and society, see *Civilization and Its Discontents*, pp. 97 f.

168. *Neurosis and Human Growth*, pp. 377–378.

169. Jurgen Ruesch, "Science, Behavior, and Psychotherapy," in Leon Salzman and Jules Masserman (eds.), *Modern Concepts of Psychoanalysis* (New York: Citadel, 1962), p. 92.

170. Fromm, *The Sane Society* (New York: Rinehart, 1955), p. 33.

171. Patrick Mullahy (ed.), *The Contributions of Harry Stack Sullivan* (New York: Hermitage, 1952), p. 121.

172. See Sullivan's description of the interview in Mullahy, *ibid.*, p. 122.

173. Alexander Gralnick, "In-Patient Psychoanalytic Psychotherapy of Schizophrenia: Problem Areas and Perspectives," in Jules H. Masserman (ed.), *Psychoanalysis and Human Values* (New York: Grune & Stratton, 1960), p. 290.

174. A psychiatric dictionary defines countertransference as follows: "When the analyst responds to the patient's interests through a stirring up of his own repressed inclinations countertransference is established." See Don D. Jackson, "Countertransference and Psychotherapy," in *Progress in Psychotherapy: 1956*, p. 235.

175. See the slashing attack on the transference neurosis by Masserman, "Transference: Counter and Countered—A Dialogue," in *Modern Concepts of Psychoanalysis*, pp. 160–173.

176. See the body of literature on transference discussed by Roy R. Grinker *et al.*, *Psychiatric Social Work: A Transactional Case Book* (New York: Basic Books, 1961), pp. 9 ff., 294 f.

177. Cf. Paul Schilder, *Psychotherapy* (New York: Norton, 1938); Franz Alexander, *Psychoanalysis and Psychotherapy* (New York: Norton, 1956); Sandor Rado, "Achieving Self-Reliant Treatment Behavior: Therapeutic Motivations and Therapeutic Techniques," in *Psychoanalysis and Human Values*, pp. 271 ff.; Walter Bromberg, *Man Above Humanity: A History of Psychotherapy* (Philadelphia: Lippincott, 1954), chap. 9.

178. See *Psychoanalysis and Psychotherapy: Selected Papers of Frieda Fromm-Reichmann* (Chicago: University of Chicago Press, 1959),

pp. 28 ff., 96 ff., *passim.* Note Edith Weigert's introduction to this book, especially p. viii.

179. See the discussion and comments by various psychiatrists (notably Russell Munroe, Judd Marmor and Frederick A. Weiss) in *Psychoanalysis and Human Values,* pp. 352 ff.

180. See *Modern Concepts of Psychoanalysis,* articles by Millett, Kelman, Masserman, Weigert, Marmor, *passim.*

181. Ludwig Binswanger, "Existential Analysis and Psychotherapy," in *Progress in Psychotherapy: 1956,* p. 148.

182. Cf. Justus Streller, *Jean-Paul Sartre: To Freedom Condemned* (New York: Wisdom Library, 1960).

183. Maurice S. Friedman, *Martin Buber: The Life of Dialogue* (New York: Harper Torchbooks, 1960), pp. 184 ff.

184. Paul E. Pfuetze, *Self, Society, Existence* (New York: Harper Torchbooks, 1961), subtitled "Human Nature and Dialogue in the Thought of George Herbert Mead and Martin Buber."

185. Hanna Colm, "Healing as Participation: Comments Based on Paul Tillich's Existential Philosophy," *Psychiatry* (1953), Vol. XVI, pp. 99–111.

186. Hazel Barnes, *The Literature of Possibility* (Lincoln, Nebr.: University of Nebraska Press, 1959).

187. Lucia Radl, "Existential and Adlerian Psychology," in *Essays in Individual Psychology,* pp. 157 ff. Cf. William Barrett, *Irrational Man* (New York: Doubleday, 1958), pp. 229–230.

188. See Frederick J. Hacker, "Freud, Marx, and Kierkegaard," in *Freud and the Twentieth Century,* pp. 125 ff. Cf. Ernst Breisach, *Introduction to Modern Existentialism* (New York: Grove Press, 1962), p. 56.

189. See Friedman, *Martin Buber,* pp. 190 ff.

190. Will Herberg, *The Writings of Martin Buber* (New York: Meridian Books, 1956), p. 14.

191. Friedman, *Martin Buber,* p. 190.

192. Quoted in *ibid.,* p. 197.

193. Binswanger, "Existential Analysis and Psychotherapy," p. 146.

194. Friedman, *Martin Buber,* p. 186.

195. Others influenced by Buber are listed in Friedman, *ibid.,* p. 162n.

196. Gabriel Marcel, *The Mystery of Being* (Chicago: Gateway Edition, 1960), Vol. II, pp. 8–9.

197. *Ibid.*

198. Paul Tillich, *The Courage to Be* (New Haven: Yale University Press, 1959), p. 165.

199. *Ibid.*, p. 124.

200. Rollo May, *Existence: A New Dimension in Psychiatry and Psychology* (New York: Basic Books, 1958), p. 11.

201. Martin Buber, *Between Man and Man* (Boston: Beacon Press, 1955), p. 204.

202. *The Courage to Be*, p. 124.

203. Jean-Paul Sartre, *Existentialism and Human Emotions* (New York: Wisdom Library, 1957), p. 53.

204. Plato, *Phaedo*, in *Great Dialogues of Plato* (New York: Mentor Books, 1956).

CHAPTER VII *The Human Image*

1. R. G. Collingwood, *The Idea of Nature* (New York: Oxford University Press, 1960 edition), pp. 3–4. Cf. F. Waismann, "The Decline and Fall of Causality," in A. C. Crombie (ed.), *Turning Points in Physics* (New York: Harper Torchbooks, 1959), pp. 84–154.

2. Cf. Gordon D. Kaufman, *Relativism, Knowledge, and Faith* (Chicago: University of Chicago Press, 1960), p. 38.

3. Michael Argyle, *The Scientific Study of Human Behavior* (London: Methuen, 1957), pp. 85–86. Cf. Gilbert Ryle, *The Concept of Mind* (London: Hutchinson, 1949), for the classic statement of this view.

4. Joseph Tussman, *Obligation and the Body Politic* (New York: Oxford University Press, 1960), p. 13. Much the same point was insisted upon, over a generation ago, by Warner Fite. "Conceiving yourself as agent the only conceivable ground of action is, never a cause, but a *reason*. . . . Accordingly, where the observer looks for causes the agent expects to find reasons. What the observer views as a relation of cause and effect is for the agent a relation of ground and consequence. For the observer the moving term is a blind force, for the agent it is a conception of value. In a word, the observer's view is mechanical, the agent's is logical and teleological." Warner Fite, *The Living Mind* (New York: Dial Press, 1930), p. 34. Fite's entire book is an examination of the significance of this view of consciousness and choice.

5. Cf. Fred Kort, "Predicting Supreme Court Decisions Mathematically: A Quantitative Analysis of the 'Right to Counsel' Cases," *American*

Political Science Review, Vol. LI (March, 1957), pp. 1–12; Franklin M. Fisher, "The Mathematical Analysis of Supreme Court Decisions: The Use and Abuse of Quantitative Methods," *American Political Science Review*, Vol. LII (June, 1958), pp. 321–338; Stuart Nagle, "Using Simple Calculations to Predict Judicial Decisions," *The American Behavioral Scientist*, Vol. IV (December, 1960), pp. 24–28.

6. For a criticism of the anthropomorphic metaphors by which advocates of mechanical (cybernetic) models in psychology and elsewhere are led to the notion of "thinking machines," see Errol E. Harris, "Mind and Mechanical Models," in Jordan Scher (ed.), *Theories of the Mind* (New York: Free Press, 1962), pp. 464–489. An ironic tendency is that these mental mechanists seek simultaneously to define the human mind as a machine (hence devoid of nonmechanical purpose) and to define the machine as potentially capable of "purpose" as well as of "consciousness."

7. See the discussion of the various meanings of "rationality" in Karl Mannheim, *Man and Society in an Age of Reconstruction* (New York: Harcourt, Brace, 1950), pp. 51 ff. The spectre of impending destruction of the "organic" or "instinctive" side of human life by the juggernaut of mechanical organization and "rationalization" is adumbrated in Roderick Seidenberg, *Post-Historic Man* (Boston: Beacon Press, 1957 edition), esp. pp. 133 ff.

8. Peter F. Drucker, *The End of Economic Man* (New York: John Day Co., 1939). Cf. Joseph A. Schumpeter, *Capitalism, Socialism, and Democracy* (New York: Harper, 1942), Part II.

9. "The worker's behavior must be understood as a response to the situation in which *the worker* believes he is placed, not to the situation as it might be conceived by an omniscient and scientifically-trained observer. One must consider the objectives which the worker is pursuing, not the objectives which the observer might consider reasonable. One must consider what the worker knows and believes about job opportunities; unknown opportunities are irrelevant. One must consider that the worker's training in self-advancement is pragmatic rather than theoretical. . . . We shall argue only that worker behavior is more nearly rational than it seems when viewed in the light of economic theory." Lloyd G. Reynolds and Joseph Shister, *Job Horizons: A Study of Job Satisfaction and Labor Mobility*, Yale Labor and Management Series (New York: Harper, 1949), p. 81. Cf. Chris Argyris, *An Introduction to Interaction Theory and Field Theory* (New Haven: Yale Labor-Management Center, 1952).

10. Charles M. Solley and Gardner Murphy, *Development of the Perceptual World* (New York: Basic Books, 1960), p. 290. For discussion see Chap. VI, above.

11. P. W. Bridgman, *The Logic of Modern Physics* (New York: Macmillan, 1927). Cf. Anatol Rapoport, *Operational Philosophy* (New York: Harper, 1954).

12. See P. W. Bridgman, *The Way Things Are* (New York: Viking Press, 1961 edition), Introduction, pp. 1–12.

13. "Attention to activities and the first person emphasizes the insight that we never get away from ourselves. . . . The problem of how to deal with the insight that we never get away from ourselves is perhaps the most important problem before us. It is associated with, but incomparably more complicated than, the problem of the role of the observer to which quantum theory has devoted so much attention and regards as so fundamental." *Ibid.*, p. 6. Cf. Bridgman, *Reflections of a Physicist* (New York: Philosophical Library, 1950), pp. 372–373.

14. See *The Way Things Are*, p. 4.

15. "It is a commonplace that we can never know anything about anything without getting into some sort of connection with it, either direct or indirect. . . . This means that no knowledge of any physical property or even mere existence is possible without interaction; in fact these terms have no meaning apart from interaction. . . . The participation of the individual is necessary in every process of intelligence, not merely in the processes of science. Intelligence can be given a meaning only in terms of the individual." Bridgman, *Reflections of a Physicist*, pp. 94–95, 373. Cf. Bridgman, *The Nature of Physical Theory* (New York: Dover, n.d. [orig. ed. 1936]), pp. 121–122; Bridgman, *The Way Things Are*, chap. 1, 7, 8, *passim*.

16. "Let me be challenging. The much flaunted exactitude of methods in the natural sciences,—is it not perhaps simply a substitute for the peculiar insight that the student of human affairs can count upon? Is not the social scientist lucky in that he is himself one of the atoms, so that the ways of atoms are familiar to him? . . . In other words, your student of human affairs can not only give a general descriptive formula, but he can also *understand* the aberrations from such recurrent phenomena quite readily . . . it is thoughtless, indeed, to deprive ourselves in the social sciences of the invaluable aid which mutual human understanding can give us, merely because the natural sciences have had to evolve techniques for getting along without it." Carl J. Friedrich, *Constitutional Government and Politics* (New York: Harper, 1937), p. 6.

17. "The form of the Platonic dialogue was quite certainly created by a historical fact—the fact that Socrates taught by question and answer. He held that form of dialogue to be the original pattern of philosophic thought, and the only way for two people to reach an understanding on any subject. And the aim of his life was to reach understanding with the people he talked to." Werner Jaeger, *Paideia: The Ideals of Greek Culture* (New York: Oxford University Press, 1943), Vol. II, p. 19. On Buber, see Chap. VI, above.

18. See Hajo Holborn, "William Dilthey and the Critique of Historical Reason," *Journal of the History of Ideas*, Vol. XI (January, 1950), pp. 116–117; H. A. Hodges, *The Philosophy of Wilhelm Dilthey* (London:

1952), *passim;* H. P. Rickman (ed.), *Wilhelm Dilthey: Pattern and Meaning in History* (New York: Harper Torchbooks, 1961), pp. 18–21; H. Stuart Hughes, *Consciousness and Society* (New York: Vintage Books, 1961), pp. 192–200; Albert Salomon, "German Sociology," in Georges Gurvitch and W. E. Moore (eds.), *Twentieth Century Sociology* (New York: Philosophical Library, 1945), pp. 590–592.

19. See the excellent summary by Rickman, *Pattern and Meaning in History,* pp. 37 ff.

20. *Ibid.,* p. 67. See also pp. 113, 120.

21. See Floyd W. Matson, "History as Art: The Psychological-Romantic View," *Journal of the History of Ideas,* Vol. XVIII (April, 1957), pp. 270–279. Cf. Hans Meyerhoff (ed.), *The Philosophy of History in Our Time* (Garden City, N. Y.: Doubleday Anchor Books, 1959), Introduction, p. 18, *passim.*

22. Cf. the acute discussion by Fritz Stern (ed.), *The Varieties of History* (New York: Meridian Books, 1956), Introduction. And see the essay by Richard Hofstadter in the same volume, "History and the Social Sciences," pp. 359–370.

23. Cf. William Barrett, *Irrational Man* (New York: Doubleday Anchor Books, 1958), p. 10.

24. See the introduction by Hans Meyerhoff to his translation of Max Scheler, *Man's Place in Nature* (New York: Noonday Press, 1962). Cf. Anna-Teresa Tymieniecka, *Phenomenology and Science in Contemporary European Thought* (New York: Noonday Press, 1962), pp. 49 ff.

25. *The Nature of Sympathy* (London: Routledge & Kegan Paul, 1954).

26. Cf. Hughes, *Consciousness and Society,* pp. 15f., 187f., 310–312, *passim.* A highly original discussion of the background of this viewpoint is to be found in Karl Stern, *The Third Revolution* (New York: Harcourt, Brace, 1954), pp. 150 ff. See also Robert S. Woodworth, *Contemporary Schools of Psychology* (New York: Ronald Press, revised edition, 1948), "The 'Understanding' Psychology," pp. 249–252.

27. Of Weber's two main statements on this theme, only one is available in English: Chapter 1 of Weber's *Theory of Social and Economic Organization,* translated by Talcott Parsons (New York: Oxford University Press, 1947). The other, an earlier and less elaborate essay, is "Über einige Kategorien der verstehenden Soziologie," published in *Logos,* Vol. IV (1913). Also useful are the three essays—especially " 'Objectivity' in Social Science and Social Policy"—published in Edward A. Shils and Henry A. Finch (eds.), *Max Weber on the Methodology of the Social Sciences* (Glencoe, Ill.: Free Press, 1949). Cf. Hughes, *Consciousness and Society,* pp. 310 ff.; Emory Bogardus, *The Development of Social Thought* (New

York: Longmans, Green, 1940), chap. 31, "Weber and Social Understanding." On James, see below, Note 35.

28. Louis Wirth, Preface to Karl Mannheim, *Ideology and Utopia* (New York: Harcourt, Brace, 1949), p. xxii. Wirth's essay is itself a vigorous and illuminating argument for the social scientist as participant-observer —"that in the realm of the social the observer is part of the observed and hence has a personal stake in the subject of observation." *Ibid.*, p. xxiv.

29. *Ibid.*, pp. 39–40.

30. A classic expression of this viewpoint is to be found in Barthold Niebuhr's testimony to the influence of the Napoleonic wars upon his history of Rome: "When a historian is reviving former times, the interest in them and sympathy with them will be the deeper, the greater the events he has witnessed with a bleeding or a rejoicing heart. His feelings are moved by justice and injustice, by wisdom or folly, by coming or departing greatness, as if all were going on before his eyes; and when he is thus moved his lips speak, although Hecuba is nothing to the player." Quoted in Emery Neff, *The Poetry of History* (New York: Columbia University Press, 1947), p. 3.

31. *Ideology and Utopia*, p. 42.

32. See Mannheim's discussion of the effort to understand the ethics of early Christian communities, with its demand for sympathetic evaluation of "the resentment of oppressed strata." *Ibid.*, pp. 40–41.

33. *Ibid.*, p. 100.

34. *Ibid.*, pp. 235–236.

35. *Ibid.*, pp. xx–xxii. See the discussion in the present work, Chap. VI, of the transactional school in perception and its debt to James, Dewey and Mead.

36. William James, *Psychology* (Cleveland and New York: World, 1948 edition), p. 14. For a more trenchant statement, cf. James, *The Varieties of Religious Experience* (New York: Mentor Books, 1958 edition), pp. 376 ff.

37. Paul E. Pfuetze, *Self, Society, Existence* (New York: Harper Torchbooks, 1961). The book was originally published in 1954 by Bookman Associates under the title *The Social Self*.

38. William Ernest Hocking, *Types of Philosophy* (New York: Scribner's, 3rd edition, 1959), Part IV, esp. chap. 39, 40. Cf. Hocking, *Human Nature and Its Remaking* (New Haven: Yale University Press, 1918).

39. *Types of Philosophy*, p. 309.

40. Alfred North Whitehead, *Modes of Thought* (New York: Capricorn Books, 1958 edition), p. 63. For a more developed statement, see

his chapter on "The Romantic Reaction," in *Science and the Modern World* (New York: Mentor Books, 1948 edition), pp. 75–96. An excellent treatment of Whitehead's "philosophy of feeling" is to be found in Albert William Levi, *Philosophy and the Modern World* (Bloomington: Indiana University Press, 1959), chap. 12.

41. *Modes of Thought*, p. 63.

42. *Ibid.*, p. 185.

43. Thus a political behavioralist declares: "Assuming that the behavior laws relating individual activity (neural, glandular, muscular, etc.) with the specified environment are known, it is possible to define the path of behavior that will be observed in the individual organism." James G. March, "An Introduction to the Theory and Measurement of Influence," *American Political Science Review*, Vol. XLIX (1955), p. 431. See the cogent critique of this and similar statements by Lee Cameron McDonald, "Voegelin and the Positivists: A New Science of Politics?" *Midwest Journal of Political Science*, Vol. I (November, 1957), pp. 233–251. McDonald's article represents a valuable effort to "encourage a perceptual balance between the objective and subjective dimensions of man the political animal. Such an achievement," he concludes, "is essential for the constructive reorientation of our sorely divided craft." *Ibid.*, p. 251.

44. Cf. Warner Fite, *The Living Mind*, chap. 2, "The Agent and the Observer," pp. 24–56. See also John Macmurray, *The Self as Agent* (New York: Harper, 1957). An eloquent expression of an equivalent point of view is set forth in Henry S. Kariel, *The Decline of American Pluralism* (Stanford: Stanford University Press, 1961), chap. 16, "An Orientation for Research," pp. 292–300.

Matter, 38, 135, 137
and motion, 29, 38
Maxwell, Clerk, 131, 132, 133, 134, 135, 138, 139, 303*n*
May, Rollo, 241, 249, 339*n*
Mayo, Elton, 56
Mead, George H., 62, 185-186, 227, 236, 256, 278*n*, 312*n*, 326*n*
Meaning, 55, 101, 248
Meaning of Evolution, The (Simpson), 172
Measurement, 23; *see also* Quantification
Mechanism, 19-126 *passim*
Cartesian, 22-26, 33, 34-35, 48, 161
and Darwinism, 36-45, 118, 119, 169-170
decline of, 131-138, 151, 161, 193, 194, 198-205
and Enlightenment, 26-31
images (and models) of, 58, 125, 130, 131, 138, 142, 157, 191, 247
Newtonian, 19-22, 37, 47, 131, 133-134, 156
in physics, 129-135 *passim*, 140, 143, 146, 150, 151
in psychology, 46-81 *passim*, 175, 179, 193, 194-205 *passim*, 220
in social scientism 31-36, 82-126 *passim*
"Mechanomorphism," 71, 244, 259
Medawar, P. B., 171, 322*n*
Meditations (Descartes), 23
Meister, D. E., 326*n*
Menninger, Karl, 221
Mentalism, 55, 70
"Mentation," 162
Merriam, Charles, 114
Metaphysics, 23, 64, 68, 109, 110
Metapsychology, 197
Meyer, Adolf, 177, 324*n*
Meyerhoff, Hans, 342*n*
Microphysics, 143, 158
Mill, James, 34, 49, 123
Mills, C. Wright, 85, 96, 252, 286*n*, 287*n*, 289*n*, 291*n*, 299*n*
Mind, 23, 29, 31, 38, 44, 116, 175-176, 177, 179-180, 215
and behaviorism, 50-51, 60, 174
and Freud, 195
"Minding," 174, 179, 226
Modes of Thought (Whitehead), 258
Mommsen, Theodor, 251
Montagu, Ashley, 170, 322*n*
Montesquieu, C. L. de Secondat de, 28, 123

Moore, Barrington Jr., 286*n*
Morality, *see* Ethics
Morene, J. L., 332*n*
Moreno, Jacob L., 223
Morgan, Lloyd, 53
Morgenthau, Hans, 86, 271*n*, 286*n*, 287*n*
Motion, 22, 23, 29, 38, 43, 48
Motivation, 76, 182, 183, 225-226
Mullahy, Patrick, 330*n*, 332*n*, 336*n*, 337*n*
Muller, Herbert J., 316*n*, 318*n*
Muller-Freienfels, Richard, 326*n*
Mumford, Lewis, 26, 178, 268*n*, 269*n*
Munroe, Ruth, 220, 335*n*, 336*n*
Murphy, Gardner, 53, 249, 273*n*, 275*n*, 276*n*, 278*n*, 327*n*, 328*n*, 340*n*
Myrdal, Gunnar, 93, 94, 289*n*
Mysticism, 24, 161, 213

Nagle, Stuart, 340*n*
Napoleon, 31, 33
Naturalism, 24, 34, 180
"Natural philosophy," 28
Nature, 31, 34, 44, 56, 133, 138, 143, 144, 243
and economics, 34-35
and man, 38-39
and Enlightenment, 33
Nature of Sympathy (Scheler), 252
Needham, Joseph, 267*n*, 270*n*, 302*n*
Neff, Emery, 272*n*, 303*n*, 343*n*
Negley, Glenn, 283*n*
Nervous system, 70, 178, 196
Neumann, J. von, 313*n*, 315*n*
Neurosis, 206, 225, 227, 230, 233
"Neutral competence," 97-99
"New Atlantis," 33
Newtonianism, 19-22, 29, 31-32, 33, 34, 156, 254
and new physics, 129, 131-135, 140, 143, 146, 150
and psychology, 47-48
religion of, 32-33
Newton, Issac, 19-22, 28, 30, 31, 36, 41, 42, 43, 66, 121, 140, 258
Niebuhr, Reinhold, 201, 331*n*
Northrop, F. S. C., 317*n*
Nuclear theory, 139-140, 141, 143, 147, 148, 158
Number, 26, 66

Objectivism, 63, 64, 66-67, 68, 150-151, 180
Observation, *see* Empiricism
Observations on Man (Hartley), 30